THE U.S. NAVY'S
SECRET SPACE PROGRAM
& NORDIC EXTRATERRESTRIAL ALLIANCE

ALSO BY MICHAEL E. SALLA, PH.D

Insiders Reveal Secret Space Programs & Extraterrestrial Alliances

BOOK ONE OF THE SECRET SPACE PROGRAMS SERIES

Kennedy's Last Stand:

Eisenhower, UFOs, MJ-12 & JFK's Assassination

Galactic Diplomacy

Getting to Yes with ET

Exposing U.S. Government Policies on Extraterrestrial Life

Exopolitics:

Political Implications of the Extraterrestrial Presence

Available at:
Exopolitics.org
Amazon.com

THE U.S. NAVY'S SECRET SPACE PROGRAM & NORDIC EXTRATERRESTRIAL ALLIANCE

Michael E. Salla, Ph.D.

Hawaii, USA

The U.S. Navy's Secret Space Program
& Nordic Extraterrestrial Alliance

Published by Exopolitics Consultants
RR2 Box 4876
Pahoa, HI 96778 USA

ISBN-13: 978-0-9986038-0-3

Library of Congress Control Number: 2017931426

Printed in the USA

Website: www.Exopolitics.Org

CONTENTS

TABLE OF FIGURES

ACKNOWLEDGEMENTS

This book would not have been possible without the testimony and documents provided by William (Bill) Tompkins, acquired over numerous phone interviews and three visits to his San Diego home. He offered invaluable suggestions on my initial draft, and was always ready to answer my questions to bring this book project to fruition. Bill is the epitome of the wise elder helping humanity transition to a "post-disclosure" world. It has been a privilege and honor to work with him, and he has my sincere respect. Therefore, I am delighted to dedicate this book to Bill Tompkins.

I am also deeply indebted to Dr. Robert Wood for first introducing me to Bill, then verifying key elements within his testimony, sharing critical documents, making suggestions on my initial book draft, and finally for writing the Foreword presented here. He is another wise elder and it has been a pleasure working with him on this project, and also on one of my earlier books where his authentication expertise on documents was invaluable regarding my research into the Kennedy Assassination.

My deep appreciation also extends to Corey Goode for his assistance and enthusiastic support of this book project. His personal briefings expanded my understanding of recent developments concerning secret space programs, and the significance of the 2016 Presidential elections for the "full disclosure" of such programs.

I wish to thank the following individuals for their contributions to the cover art: Daniel Gish for his spaceship graphic, representing Corey Goode's description of the Arnold Sommerfeld science research vessel belonging to the Solar Warden program; "Luc" for permission to use the fractal "The Awakening III Rebirth © 20XX CygX1" (cygx1.deviantart.com) on the back cover; Xavier

Hernon for permission to use the fractal "Strange" (hmn.deviantart.com) on the front cover; and special thanks to Rene McCann for her talent and graphics expertise in adding final touches and special effects to the cover art to enhance its overall quality.

Many thanks go to Duke Brickhouse for his help in filing the Freedom of Information Act requests that led to the obtainment of 1500 pages of documents concerning Rear Admiral Rico Botto, which were invaluable for authenticating William Tompkins' testimony.

I am very grateful to Rear Admiral Larry Marsh (USN, ret.), Captain Larry Boeck (USN, ret.), and Commander Art Lumley (USN, ret.) for answering my questions, and for their willingness to go on public record about their knowledge of William Tompkins and his activities.

My thanks go to Gaia TV for permission to use the graphic illustration of the hybrid air/spacecraft used in multiple abduction incidents involving Corey Goode.

Much appreciation goes to A. Hughes for copy editing the final manuscript, and assisting once again in my book projects.

Finally, I am deeply grateful to my dear wife and soulmate, Angelika Whitecliff, who cheerfully took on the indispensable role of chief editor for this book. She also designed the cover art, and formatted the final version you have before you. I am truly blessed to have such a wonderful human being who supports my writing and truth telling endeavors in every possible way.

Michael E. Salla, Ph.D.
February 23, 2017

FOREWORD

When Michael gave me an advance copy of his book (the one you are about to read), I turned the pages with great interest because his most recent book was only two years ago, (*Insiders Reveal…Secret Space Programs and Extraterrestrial Alliances ©2015*). What amazed me in this current book was his ability to create a highly credible scenario of how the innermost secrets of UFO history have been kept from the public and how the Navy may have been able to create a secret space program of its own that the Air Force and CIA might not even know about.

One of the reasons I was grateful to be asked to write this foreword is that Michael Salla's earlier work *(Kennedy's Last Stand)* took great advantage in referring as evidence the questioned "leaked" documents that my son Ryan and I put on the website, www.majesticdocuments.com. Those documents and their apparent authenticity lead me to listen carefully when I first meet people with wild personal stories. Dr. Salla takes the unfolding testimony of a source who approached me, William M. Tompkins, and synthesizes his information with other key witnesses such as Corey Goode, showing that many of the details of their reports are exceptionally consistent. Examples would include development of the Nazi capability in Antarctica before, during and after WWII and their relationship with alien races.

This book is chronologically comfortable, beginning correctly with the "Battle of L.A." as a critical event in triggering an intense interest in Government study of the UFO dilemma and linking this to the William Tompkins early involvement with both UFOs and the Navy. His remarkable assignment was to listen to the briefings of Navy spies sponsored creatively by James Forrestal before the US entry into the war. Many of the recollections of what these spies said was going on in Germany and Antarctica are then correlated with the reports from other sources and used to confirm the deep awareness of the Navy in the details of the Nazi technology sophistication.

Salla highlights the specific testimony in the Tompkins autobiography, *Selected by Extraterrestrials*, (of which I was editor). This book provides personalized details of his involvement in the Douglas "think tank." Tompkins reports the decision to create the RAND Corporation specifically to study the alien ET problem. Salla properly emphasizes the assignment of Tompkins the draftsman to draw designs of kilometer-long planned Naval spacecraft in the 1950's. They were presumably precursors to the Navy's later Solar Warden Program, now in space since the 1980s.

Many of us have heard the rumors that somehow the *Star Trek* television program might have been influenced by reality, but Salla lays out the specific details of the links of the executive producer, Gene Roddenberry, to Vice Admiral Leslie Stevens III. He was the possible source of some of the themes based on his involvement in leaking the plots related to his involvement in psychological warfare. Some of the stories may have been based on the facts of some alien interactions. The reality of what we encountered in our covert studies may have told us who the good guys were (Nordics = Capt. Kirk) and the bad guys (Draco Reptilians = Klingons).

This book is consistent with Salla's other works in having superb references to nearly every claim or unusual allegation. His hundreds of references are normally to written works, but if his source is electronic, he provides whatever electronic source he had at the time. I checked several of them, and they are still alive with content as I write this foreword.

One of the most stimulating aspects of this book is Salla's willingness to speculate about the relations that may have affected the Trump election results and how the FBI and maybe even the Nordics could have played a role. He makes it clear that in his judgment there is evidence that would support the contention that the CIA has been infiltrated by the Nazis and Reptilians, and that we may right now be on the threshold of seeing whether the good guys (US Navy, FBI and the Nordics) vs. bad guys (CIA, USAF, NSA and the Cabal) will win and define the future of our world. He notes that Trump has selected military advisors who may be familiar with the Secret Space Programs and have declared a plan to significantly increase the Navy budget.

One of the more impressive results reflected in this work is Salla's reliability to verify and validate. For example, when Tompkins claimed that he had worked for Rick O'Botta, (as spelled in the first edition of Tompkins' book), he determined that he was working for Rico Botta based on copies of his signature, and then researched Botta's Naval record in the archives, showing that he was indeed the person assigned to be in charge of the Naval Air Station at the time of Tompkins' assignment there. This example is a powerful reason to accept the claims in this book as being supported by evidence.

Another thing that impresses me is that Michael Salla could extract exceptional insight to the reports of William Tompkins from personal conversations as well as a scrutiny of some of his interviews with radio show host Jeff Rense, learning many things that I, as Tompkins' close collaborator and book editor, had never heard of. He has woven these facts seamlessly into the story of the remarkable complicity of different organizations that created secret space programs unknown to the public.

My own personal involvement in the things that led to this book is covered beautifully in the Preface that follows. However, although Salla's last book came out in 2015, I didn't get a copy until early 2016 after Tompkins' book was already out. I ordered a copy from Amazon and had it mailed directly to Tompkins. As noted in the Preface, he was stunned, because he thought that he was going to be the first person to reveal the "role of Nazi Germany and the US Navy in developing secret space programs."

The development of the secret space programs would have been dependent on advanced technology, which is not a secret to those who built the craft. Our scientific knowledge from aliens must transcend anything in the open literature. I would have welcomed some reference to what Salla's views might be on time travel or "jump rooms" that allegedly have permitted us to go to Mars and other solar systems not using craft. Maybe that will be in his next book. In the meantime, read this one.

By Robert M. Wood
Ph.D. in Physics, Cornell University, 1953.
Research and Development Manager,
1949-1993 Douglas/McDonnell Douglas.
UFO Investigator and author 1994-present.

PREFACE

In December 2015, a remarkable book was published detailing a highly credentialed aerospace engineer's direct exposure to secret space programs and extraterrestrial intervention in human affairs. What immediately distinguished William Tompkins' book, *Selected by Extraterrestrials: My life in the top secret world of UFO's, think tanks and Nordic secretaries*, was the extensive number of documents he presented that substantiated key aspects of his testimony. The documents he included could be confirmed and fact checked. Information found within them allowed for further investigation, especially in terms of historic records and Freedom of Information Act documents, to confirm whether or not he was revealing actual historic events that had taken place.

What lent even greater credibility to Tompkins' book was that its editor, Dr. Robert Wood, had worked at the same aviation company where Tompkins says he was asked to design kilometers-long secret spacecraft for the US Navy. Dr. Wood was employed with Douglas Aircraft Company (later McDonnell Douglas) over a 43 year period (1949-1993), and worked there at the same time as Tompkins (1950-1963), even though the two had never met prior to 2009.[1]

Among Dr. Wood's assignments while at Douglas was to research UFO reports to determine the feasibility of flying saucer designs for the aerospace industry. He was assigned this task by Douglas after he met with chief executives:

> One day after I reported to a couple of VP's on how we were doing, one of them asked me personally if I was doing anything interesting outside of my job. 'You're not going to believe this, but I've read about 50 books on UFOs', I said, and 'the amazing conclusion I have come to is that they are very real extraterrestrial craft'. The only thing that's

> uncertain is whether we find out how they work before or after our competitor Lockheed. After a moment of silence one of them (VP) said, 'how much would it cost to take a look at that question'? Therefore we started a project, quite low key, to take a look at the question of how they work.[2]

Dr. Wood was able to corroborate the names of key engineers, scientists and projects at Douglas, which Tompkins referred to; especially senior company officials such as Elmer Wheaton and Dr. Klemperer who are key figures in what Tompkins said took place in a secret think tank at Douglas. Regarding this Dr. Wood wrote:

> I knew or had met nearly all of the people he referenced, especially the key persons he worked for in a "think tank", Elmer Wheaton and his German scientific advisor, Wolfgang B. Klemperer. His story of the work he did exactly coincides with what I remember, although I was not aware that there was a think tank at that time at that location. [3]

Dr. Wood was so impressed with Tompkins' detailed and accurate testimony that he decided to assist him by becoming the editor of his autobiography. He wrote the following conclusion in his Preface to Tompkins' book:

> I have total confidence that he is honestly telling the story as he best remembers it. Fortunately, Bill had kept some copies of some photos and documents that support this story and are included as figures as they apply. They do indeed enhance the credibility of this man's amazing life so far.[4]

In January 2016, after I had heard several interviews featuring both Tompkins and Dr. Wood, which aired in December 2015 on the popular *Jeff Rense* radio show, I immediately got in touch with Dr. Wood to learn more about Tompkins.[5] He vouched for Tompkins bona fides and the significance of his testimony for understanding the UFO phenomenon. What I found especially intriguing was that Dr. Wood pointed out that when he read my own book, *Insiders Reveal Secret Space Programs and Extraterrestrial Life*, which had been published in September 2015, he was struck by the similarities with Tompkins' information. He immediately ordered a copy for Tompkins and gave it to him. Dr. Wood said that Tompkins reaction upon reading my own book was one of astonishment. I subsequently spoke by phone with Tompkins. He stated that he had been told that he would be the first to reveal the role of Nazi Germany and the US Navy in developing secret space programs, and the role played by different extraterrestrial civilizations.

My book investigated the whistleblower claims of Corey Goode, who said he served in a secret space program created by the Navy called Solar Warden, as well as other related programs for a twenty year period, from 1987 to 2007. I found much in the way of historic documents, additional whistleblower testimony and circumstantial evidence which supported Goode's startling claims. Tompkins' information added significant details and, most importantly, documents that helped validate key parts of Goode's testimony. More significantly, as you will read in chapter nine of this book, there is a very good reason why there are so many similarities between Tompkins and Goode's accounts of respective Nazi Germany and US Navy secret space programs.

After speaking with Dr. Wood and Tompkins in January, I decided to travel from my Hawaii home to San Diego to meet with them both in February 2016. I encountered Tompkins as someone who despite his advanced age (92 at the time) still possessed a sharp mind and great attention for detail. While no longer having a photographic memory, which marked him out as exceptional early in his aerospace career, Tompkins could still remember the bulk of what he had experienced.

At his San Diego home, Tompkins brought out six boxes of documents which Dr. Wood and I examined. He allowed us to copy the most relevant documents for our respective investigations. These documents were invaluable since they added more details about Tompkins' long and distinguished career in the aerospace industry, and with the US Navy in various capacities. Significantly, despite his official retirement, Tompkins continues to work as a 'consultant' for the Navy with a current security clearance. I have travelled two more times to San Diego to meet with Tompkins (October and December 2016). On each visit, Dr. Wood and I have been able to see more of his documentation, which he needs to retrieve from a nearby storage area—some of which are reproduced in this book.

In mid-2016, I began working with an attorney, Duke Brickhouse, Esq., to track down Freedom of Information Act documents that shed light on Rear Admiral Rico Botta, the senior Navy officer Tompkins identified as the key figure in a World War II covert Navy operation to spy on Nazi Germany's secret aerospace programs. The FOIA documents we received (over 1500 pages) provided conclusive evidence that supported Tompkins' claims of having worked with Admiral Botta on a covert intelligence gathering project related to experimental aircraft research. Some of these documents are presented in chapter two.

Chapters three and four discuss developments at Douglas Aircraft Company and its offshoot, Project RAND, after World War II concerning intelligence briefing documents about Germany's secret space program, and the study of crashed UFOs. Of special interest is Tompkins' account of the role played by competing groups of extraterrestrials, described by him as "Reptilian" and human-looking "Nordics," in the development of advanced aerospace technologies both in Nazi Germany and the US. Tompkins says that the "Nordics" had infiltrated Douglas Aircraft, and helped him create designs for future US Navy space battle groups.

Additional FOIA requests were made concerning other Navy officers identified by Tompkins as relevant to what had occurred in Nazi Germany, and the Navy's subsequent development of a secret space program. The FOIA documents received to date provide important details, which corroborate key aspects of the big picture Tompkins has revealed about Navy space battle fleets capable of interstellar operations. Chapters five and eight will discuss these "big picture" aspects in terms of "soft disclosure," where two popular science fiction series can be directly connected to the Navy's secret space program that Tompkins helped design.

I also was able to track down and interview three retired US Navy officers (a Rear Admiral, a Captain and a Commander) who were willing to attest to Tompkins' bona fides, as an expert in areas of special interest to the Navy, when he was President of the Rogue Valley Council Navy League from 1991 to 1999. In particular, two of the officers were identified on a document provided by Tompkins revealing the existence of a "Special Projects Committee" that Tompkins says discussed plans for various extraterrestrial and secret space program related projects. Two of the Navy officers confirmed that the Special Projects Committee existed, and that Tompkins often discussed the extraterrestrial topic in meetings. Importantly, they confirmed that Tompkins was regarded as an expert on the extraterrestrial topic and was well connected to important Navy officials when it came to this subject. Documents pertinent to the projects Tompkins conducted at the Navy League are found in chapter seven.

It is important to note that Tompkins states he was permitted to come forward with his testimony and documents with the approval of a senior Navy official wanting to disclose the existence of the Navy's secret space program. Tompkins' claim that his disclosures are covertly sanctioned by the Navy is borne out by his continued work as a Navy consultant, and is confirmed within a document he confidentially showed Dr. Wood and I.

Tompkins' continued relationship with the Navy while coming forward with his testimony becomes even more momentous given the outcome of the 2016 US Presidential election. Chapter 10 reveals a limited disclosure initiative that involved key figures in the Clinton campaign. They were informally briefed about a "limited disclosure" initiative involving a US Air

Force run secret space program, which operates independently of the Navy's more technologically advanced program. Chapter 11 focuses on the key institutional relationships that President Trump needs to develop in order for "official disclosure" of secret space programs to occur. What this means for ending a technological divide between the vast majority of humanity and those belonging to a smaller breakaway human civilization is vital to understand.

The prospects of "full disclosure" of secret space programs and extraterrestrial life under President Trump's administration is examined in chapter 12, in light of his choices of senior national security personnel. The prospects of them gaining "need to know" access to secret space program information in order to competently brief Trump will be assessed, as will Trump's likely response. The final chapter takes a close look at the role of the "Nordics" and other friendly extraterrestrials in terms of what their technological and spiritual assistance means for full disclosure, as well as preparing humanity for predicted solar and cosmic events that are expected to radically transform life on Earth.

The documents presented in this book not only directly support the testimony of William Tompkins, but also corroborate the testimony of Corey Goode and other secret space program whistleblowers. This book should therefore be considered as volume two to *Insiders Reveal Secret Space Program and Extraterrestrial Alliances*. The information in these two books about the connection between the historic development of multiple secret space programs and different extraterrestrial civilizations is life changing—it certainly was for me.

Michael E. Salla, M.A., Ph.D.
Kaimu-Kalapana, Hawaii
February 7, 2017

Endnotes

[1] Dr. Wood describes his time at Douglas in the Preface to William Tompkins, *Selected by Extraterrestrials: My life in the top secret world of UFO's, think tanks and nordic secretaries*, (Createspace, 2015), p. v.
[2] Rense.com, "Dr. Robert Wood - Aerospace Engineer Veteran Blows The Whistle On UFOs," http://rense.com/general96/woodsvet.html (Accessed 3/18/16).
[3] William Tompkins, *Selected by Extraterrestrials,* p. v.
[4] William Tompkins, *Selected by Extraterrestrials,* p. vi
[5] A list of Tompkins interviews with Jeff Rense can be found here: http://exopolitics.org/william-tompkins-us-navy-secret-space-program/

CHAPTER 1

SPECTACULARLY ENIGMATIC LOS ANGELES AIR RAID

On the night of February 24, 1942 and on into the early hours of the 25th, residents of Los Angeles, California were awakened by an extraordinary UFO event. An unknown number of unidentified aircraft silently flew in a 40-mile arc across the sky, centered over Long Beach, while the US Army and Navy ships in Long Beach Harbor unleashed anti-aircraft artillery barrages against the mysterious objects. They were feared to be Japanese aircraft about to launch another surprise Pearl Harbor-like attack. Hundreds of thousands witnessed the hovering UFOs, which were well lit up for several hours by searchlights and the anti-aircraft artillery during a total blackout. The UFOs did not attack but damage was done by the artillery shells which fell to the ground causing significant property losses.

The events of that night were extensively covered by major newspapers, including the *Los Angeles Times,* which wrote a number of stories about the unexploded artillery shells and property damage. Here is how a 1983 book officially documenting US Air Force history, *The Army Air Forces in World War II*, described the incident which involved as many as 25 unknown craft flying near Long Beach as written in an Army Colonel's report:

During the night of 24/25 February 1942, unidentified objects caused a succession of alerts in southern California... An alert called at 1918 [7:18 p.m., Pacific time] was lifted at 2223, and the tension temporarily relaxed. But early in the morning of the 25th renewed activity began. Radars picked up an unidentified target 120 miles west of Los Angeles.

Figure 1. Los Angeles Times Coverage of Air Raid Incident

Los Angeles Times

Searchlights and Anti-aircraft Guns Comb Sky During Alarm

Antiaircraft batteries were alerted at 0215 and were put on Green Alert—ready to fire—a few minutes later. The AAF kept its pursuit planes on the ground, preferring to await

> indications of the scale and direction of any attack before committing its limited fighter force. Radars tracked the approaching target to within a few miles of the coast, and at 0221 the regional controller ordered a blackout. Thereafter the information center was flooded with reports of "enemy planes, " even though the mysterious object tracked in from sea seems to have vanished. At 0243, planes were reported near Long Beach, and a few minutes later a coast artillery colonel spotted "about 25 planes at 12,000 feet" over Los Angeles. At 0306 a balloon carrying a red flare was seen over Santa Monica and four batteries of anti-aircraft artillery opened fire, whereupon "the air over Los Angeles erupted like a volcano."[6]

This book, published by the Office of Air Force History, went on to describe unconfirmed reports of several of the UFOs being shot down:

> These mysterious forces dropped no bombs and, despite the fact that 1,440 rounds of antiaircraft ammunition were directed against them, suffered no losses. There were reports, to be sure, that four enemy planes had been shot down, and one was supposed to have landed in flames at a Hollywood intersection. Residents in a forty-mile arc along the coast watched from hills or rooftops as the play of guns and searchlights provided the first real drama of the war for citizens of the mainland.[7]

In contrast to the Air Force official history book's account of the Los Angeles Air Raid's mystery craft, the official explanation by Frank Knox, Secretary of the Navy, at a Press Conference on the morning of the event was that the incident was "just a false alarm" caused by jittery nerves. His statement is illustrated in the following February 25, 1942, Los Angeles Times article.

Figure 2. News Coverage of Navy Secretary's View of LA Air Raid

Knox Indicates Raid Just 'Jittery Nerves'

By a Times Staff Correspondent

WASHINGTON, Feb. 25.—Secretary of Navy Knox indicated today that this morning's reported air raid over Los Angeles area was due largely to "jittery nerves."

The Secretary when asked during his press conference concerning the Army's Western Command statement that an unidentified air raid had been reported over the West Coast area said his information indicated that it was "just a false alarm."

"There were no planes over Los Angeles last night, at least that's our understanding," the Secretary declared. "None have been found and a very wide reconnaissance has been carried out."

The Secretary then intimated that the reports might be due largely to jittery nerves.

He related that during the World War when a new outfit moved up to the front the horns used to warn gas attacks were "going all the time" for the first three or four nights, when actually no gas attack had been made.

In his discussion of various reports of enemy action along the West Coast Knox said he "suspected" that eventually vital industries in that area will have to be moved to safer inland re

Turn to Page A, Column 1

Henry Stimson, Secretary of the Army, waited a day before issuing a statement that laid the foundation for the subsequent 1983 official Air Force history of the Los Angeles incident. He said that up to five unidentified craft were involved in the incident and proposed two explanations:

> [L]ocal commanders ... indicated a belief that from one to five unidentified airplanes had been over Los Angeles. Secretary Stimson announced this conclusion as the War

> Department version of the incident, and he advanced two theories to account for the mysterious craft: either they were commercial planes operated by an enemy from secret fields in California or Mexico, or they were light planes launched from Japanese submarines. In either case, the enemy's purpose must have been to locate anti-aircraft defenses in the area or to deliver a blow at civilian morale.[8]

Figure 3. Los Angeles Times Covers Army Secretary on LA Air Raid

ALL THE NEWS ALL THE TIME

Los Angeles Times

IN THREE PARTS — 36 PAGES

TIMES OFFICE

DAILY, FIVE CENTS

ARMY SAYS ALARM REAL

U.S. Flyers Reap Indies Victories

Sink Two Transports and Destroy Three Planes; MacArthur Breaks Lull by Successful Attacks; American-British Airmen Blast Foe in Burma

Storm Grows Over Delay in Alien Ouster

INFORMATION, PLEASE

Five Deaths Laid to Raid Blackout

Roaring Guns Mark Blackout

Identity of Aircraft Veiled in Mystery; No Bombs Dropped and No Enemy Craft Hit; Civilians Report Seeing Planes and Balloon

The conflicting Army and Navy explanations served only to confuse the situation as a *New York Times* editorial explained:

> *The New York Times* on 28 February expressed a belief that the more the incident was studied, the more incredible it became: "If the batteries were firing on nothing at all, as Secretary Knox implies, it is a sign of expensive incompetence and jitters. If the batteries were firing on real planes, some of them as low as 9,000 feet, as Secretary Stimson declares, why were they completely ineffective? Why did no American planes go up to engage them, or even to identify them?[9]

Knox's quick dismissal of the Los Angeles UFO incident appeared to be an effort to cover-up the seriousness of what had happened. This was highlighted by the Army's more forthright statement about the unidentified aircraft that were involved. What was the Navy trying to cover up?

Among the thousands of witnesses to the event was a young William Tompkins, who was 17 years old at the time and living with his parents in Long Beach. He got to see the extraordinary event up close. Now at age 93, Tompkins wrote about the incident in his 2015 autobiography, *Selected by Extraterrestrials*:

> On February 25, 1942, three months after the attacks on Pearl Harbor, a very strange thing happened. By this time, my family had moved to Long Beach. We were now living in a high second-floor apartment that was converted from a large home. It was only four blocks from the ocean. At about 8:00 pm on the night in question, my father called my brother and me to our deck, which faced the bay. There was a strange intense light just above the horizon: a narrow beam pointing towards the ocean. The little beam turned horizontally, right into our eyes, and hit the back wall of our apartment and the surrounding trees. It was blinding. Suddenly, and bafflingly, the light went out. Whatever it was, had gone. There was nothing we could do but stand there amazed. Finally we retired to our beds.
>
> Just after midnight, the air-raid sirens and anti-aircraft guns of the coastal artillery woke everyone up. We ran out onto the street and saw a large round craft, above seven thousand-feet up, floating in the air above us. It slowed to a stop right overhead and remained stationary. It was lit up by eight searchlights, while anti-aircraft shells burst all around and against it. Most of the shells exploded on the bottom of the craft – we just couldn't believe the thing hadn't exploded or been shot down. Three, then five, other ships appeared near it; some of the searchlights, as

well as the anti-aircraft teams, focused on each one as they passed the first object. Eventually, the first craft slowly departed as well.[10]

Tompkins and many other reports support the official Army position that there were several unidentified aircraft involved in the Los Angeles Air Raid incident. The Los Angeles Examiner reported that civilian witnesses had estimated 50 planes that slowly traveled in a "V" formation.[11]

Some of the UFO craft stopped motionless over Long Beach, according to Tompkins. Clearly, these were not conventional aircraft, and displayed an advanced propulsion system unknown at the time. It would be after the war that the first helicopters would be built with the ability to hover, yet these hovering craft over Los Angeles were silent according to Tompkins and other witness reports. The motivation for the Navy's Secretary Knox to cover-up the incident now becomes clearer. The unknown craft's ability to hover and withstand an artillery barrage would have sent shock waves through the general public and the War Department.

There were several aircraft companies in the Los Angeles area which included the Douglas Aircraft Company, headquartered in Santa Monica, and Lockheed Aircraft, headquartered in Burbank. We know that senior personnel from Douglas, including its President Donald Douglas, personally witnessed these events, according to Tompkins, who later worked for 12 years with the Douglas Aircraft Company.[12] These aviation industry leaders immediately realized the revolutionary significance of the propulsion systems displayed by the hovering UFOs as described by Tompkins.

In addition, Long Beach Naval Shipyard at the time was under the direct flight path of the UFOs, and this would have been a major concern for the Navy. Were the UFOs part of a reconnaissance mission by Japan, or other Axis powers, to ascertain the strength of anti-aircraft defenses for a possible future attack, as the Secretary of the Army hypothesized? Or, were the UFOs interplanetary in origin raising the possibility of revolutionary propulsion systems that could radically transform the aviation industry and significantly impact the war effort?

In either case, both the Office of Naval Intelligence and Army Intelligence (G-2) would certainly have attached great urgency to studying the UFOs involved in the incident, and especially retrieving any artifacts from the craft targeted by the artillery barrage. This leads to the question of whether any of the reports of up to four craft being shot down were in any way accurate, despite the Army's official dismissal of such reports? According to the *Los Angeles Examiner*, civilians reported that three craft had been shot down while over the ocean.[13] Tompkins says that he learned from confidential Navy sources (he was working at the time as an asset for the Office of Naval Intelligence) that two flying saucer-shaped craft were shot down.[14] The saucers were recovered respectively by the Navy and the Army, and it was discovered that the saucer-shaped craft were unmanned fully automated drones.[15]

A leaked official document, whose authenticity is still disputed, refers to two UFO craft being shot down and being secretly retrieved respectively by the Navy and the Army:

> … regarding the air raid over Los Angeles it was learned by Army G2 that Rear Admiral Anderson … Naval Intelligence, has informed the War Department of a naval recovery of an unidentified airplane off the coast of California with no bearing on conventional explanation. Further it has been revealed that the Army Air Corps has also recovered a similar craft in the San Bernardino Mountains east of Los Angeles which cannot be identified as conventional aircraft. This Headquarters has come to the determination that the mystery airplanes are in fact not earthly and according to secret intelligence sources they are in all probability of interplanetary origin.[16]

The document is an alleged Top Secret Memorandum from Chief of Staff of the Army, George Marshall, to President Roosevelt, dated March 5, 1942. According to the document, Marshall:

> ... issued orders to Army G2 that a special intelligence unit be created to further investigate the phenomenon and report any significant connection between recent incidents and those collected by the director the office of Coordinator of Information. [17]

This is allegedly the genesis of the Army's legendary "Interplanetary Phenomenon Unit."

The Interplanetary Phenomenon Unit (IPU) was a highly classified army intelligence unit that did exist during the World War II era. After initially denying its existence, the US Air Force was eventually forced to acknowledge that the Interplanetary Phenomenon Unit did exist for a period of time. Documents released through the Freedom of Information Act (FOIA) confirmed the existence of this highly secretive investigatory group, despite the best efforts by Air Force officials to cast doubt on its existence.

In May 1984, for example, UFO researcher William Steinman sent a FOIA request to the Army Directorate of Counterintelligence. Steinman received a reply from a Lieutenant Colonel Lance R. Cornine:

> As you note in your letter, the so-called Interplanetary Phenomenon Unit (IPU) was disestablished and, as far as we are aware, all records, if any, were transferred to the Air Force in the late 1950's. The 'unit' was formed as an in-house project purely as an interest item for the Assistant Chief of Staff for Intelligence. It was never a 'unit' in the military sense, nor was it ever formally organized or reportable, it had no investigative function, mission or authority, and may not even have had any formal records at

Figure 4. Alleged Memorandum from General Marshall to Roosevelt

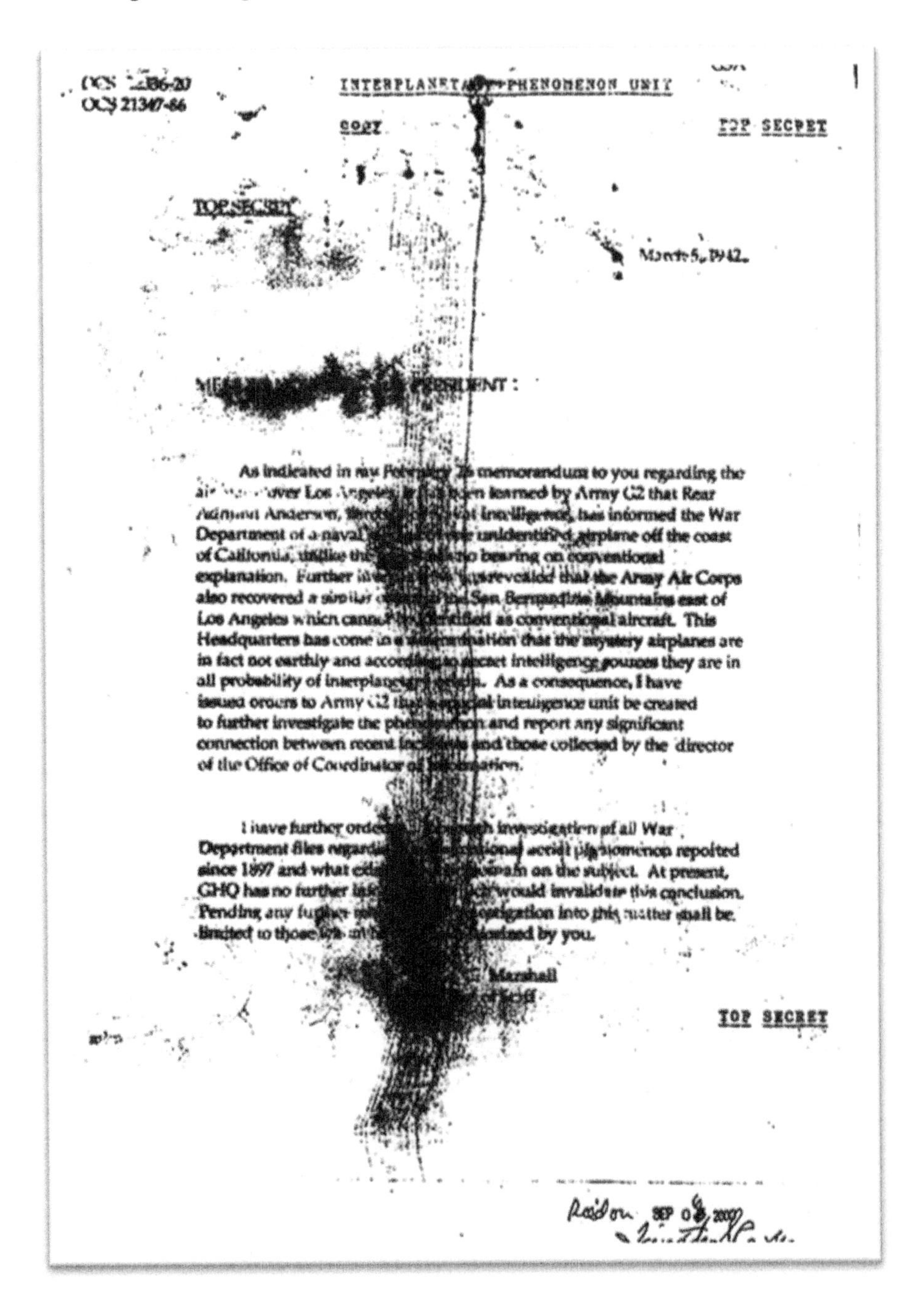

OCS [illegible]
OCS 21347-66

INTERPLANETARY PHENOMENON UNIT

COPY TOP SECRET

TOP SECRET

March 5, 1942.

MEMORANDUM FOR THE PRESIDENT:

As indicated in my February 28 memorandum to you regarding the air [illegible] over Los Angeles, it has been learned by Army G2 that Rear Admiral Anderson, [illegible] intelligence, has informed the War Department of a naval [illegible] unidentified airplane off the coast of California, [illegible] bearing on conventional explanation. Further [illegible] revealed that the Army Air Corps also recovered a similar [illegible] San Bernardino Mountains east of Los Angeles which cannot be identified as conventional aircraft. This Headquarters has come to a determination that the mystery airplanes are in fact not earthly and according to secret intelligence sources they are in all probability of interplanetary origin. As a consequence, I have issued orders to Army G2 that a special intelligence unit be created to further investigate the phenomenon and report any significant connection between recent incidents and those collected by the director of the Office of Coordinator of Information.

I have further ordered [illegible] investigation of all War Department files regarding [illegible] aerial phenomenon reported since 1897 and what [illegible] on the subject. At present, GHQ has no further [illegible] would invalidate this conclusion. Pending any further [illegible] investigation into this matter shall be limited to those [illegible] by you.

[illegible] Marshall
[illegible] of Staff

TOP SECRET

Reid on SEP 0[illegible] 2007

> all. It is only through institutional memory that any recollection exists of this unit. We are therefore unable to answer your questions as to the exact purpose of the unit, exactly when it was disestablished, or who was in command. This last would not apply in any case, as no one was in 'command'. We have no records or documentation of any kind on this unit."[18]

Cornine's letter did acknowledge that the IPU did exist, but downplayed its existence as merely an "interest item" that was never an operational army "unit" of any kind.

In March 1987, British researcher Timothy Good also sent a FOIA request to the US Army's Directorate of Intelligence. Good received a reply from Colonel William Guild. Guild's letter not only confirmed the IPU's existence, but also revealed that the IPU was operational at one time:

> Please be advised that the aforementioned Army unit was disestablished during the late 1950's and never reactivated. All records pertaining to this unit were surrendered to the U.S. Air Force Office of Special Investigations in conjunction with operation BLUEBOOK.[19]

So according to the Army's Directorate of Intelligence, the IPU did exist and was abolished in the late 1950's. No time line is given for when it was active.

Consequently, the leaked March 5, 1942 memorandum, where General Marshall ordered "Army G2" to create "a special intelligence unit" to "investigate the phenomenon,"[20] almost certainly was referring to the Interplanetary Phenomenon Unit. This is the conclusion reached by Dr. Robert Wood and his son Ryan in their respective efforts to authenticate the Marshall Memorandum:

> The memo bears correct Office of Chief of Staff (OCS) file numbers and has "Interplanetary Phenomenon Unit" (IPU)

> typed on it at a later time by a different typewriter. It is logical to believe that this is the order that sets up the IPU.[21]

Additional support for the authenticity of the leaked March 5, 1942 Marshall Memorandum comes from a document officially released in 1974, through the Freedom of Information Act.[22] The document is dated February 26, 1942, and again involved a memorandum from General Marshall to President Roosevelt. Issued one day after the Los Angeles Air Raid, it provides a preliminary report of what occurred that is consistent with the contents of the March 5 document, issued one week later. In the February 26 Memorandum, Marshall tells Roosevelt:

> The following is the information we have from GHQ at this moment regarding the air alarm over Los Angeles of yesterday morning:
>
> From details available at this hour:
>
> 1. Unidentified airplanes, other than America Army or Navy planes, were probably over Los Angeles, and were fired on by elements of the 37th CA Brigade (AA) between 3:12 and 4:15 am. These units expended 1430 rounds of ammunition.
> 2. As many as fifteen airplanes may have been involved, flying at various speeds what is officially reported as being "very slow" to as much as 2000 mph and at elevations from 9000 to 18000 feet.
> 3. No bombs were dropped.
> 4. No casualties among our troops.
> 5. No planes were shot down.
> 6. No American Army or Navy planes were in action.
>
> Investigations continuing. It seems reasonable to conclude that if unidentified airplanes were involved they may have been from commercial sources, operated by enemy agents for purposes of spreading alarm,

Figure 5. Feb 26, 1942 Memorandum from Marshall to Roosevelt

SECRET

File

OCS 21347-86

February 26, 1942.

Record Section

MEMORANDUM FOR THE PRESIDENT:

The following is the information we have from GHQ at this moment regarding the air alarm over Los Angeles of yesterday morning:

"From details available at this hour:

"1. Unidentified airplanes, other than American Army or Navy planes, were probably over Los Angeles, and were fired on by elements of the 37th CA Brigade (AA) between 3:12 and 4:15 AM. These units expended 1430 rounds of ammunition.

"2. As many as fifteen airplanes may have been involved, flying at various speeds from what is officially reported as being 'very slow' to as much as 200 MPH and at elevations from 9000 to 18000 feet.

"3. No bombs were dropped.

"4. No casualties among our troops.

"5. No planes were shot down.

"6. No American Army or Navy planes were in action.

"Investigation continuing. It seems reasonable to conclude that if unidentified airplanes were involved they may have been from commercial sources, operated by enemy agents for purposes of spreading alarm, disclosing location of antiaircraft positions, and slowing production through blackout. Such conclusion is supported by varying speed of operation and the fact that no bombs were dropped."

DECLASSIFIED
E.O. 11652, Sec. 3(E) and 5(D) or (E)
OSD letter, May 3, 1972
By [illegible] NARS Date 4-9-74

(Sgd) G. C. MARSHALL
Chief of Staff.

21347
86

akn

SECRET

Orig. dispatched to Pres. 2/27/42

> disclosing locations of antiaircraft positions, and slowing production through black out. Such conclusion is supported by varies speed of operation and the fact that no bombs were dropped.[23]

The official February 26 memorandum points out that further Army investigations were being conducted, indicating that Marshall would send updates on the incident as more was learned.

Significantly, the February 26 memorandum refers to the speed of some of the unidentified craft as "very slow," corroborating Tompkins' account that some of the craft were indeed hovering over Los Angeles. This helps authenticate that the craft were non-terrestrial in origin since the only known hovering capable aircraft at the time were balloons/dirigibles, which would have been easy targets for the anti-aircraft guns. Consequently, after the first memorandum was sent on February 26, Marshall was informed that two of the unidentified craft had been located, and possessed highly advanced propulsion technologies. These developments, after the February 26 memo was issued, would have required a follow-up memorandum by Marshall to Roosevelt, which is what the March 5 memorandum appears to be.

Finally, further support for the authenticity of the March 5, 1942 Marshall Memorandum may come from the Director of the FBI, J. Edgar Hoover. In an official memo released through the Freedom of Information Act, Hoover handwrote on July 15, 1947 (a week after the alleged Roswell UFO crash): "We must insist upon full access to disks recovered. For instance, in the La case the Army grabbed it and would not let us have it for cursory examination."[24] Was Hoover referring to a flying disc recovered after the Los Angeles Air Raid, or a disc stored at Los Alamos (La) National Laboratory after the Roswell incident? Whatever the answer, Hoover is indicating that the Army was studying recovered flying discs and not sharing information with the FBI.

So, if Army Intelligence created the Interplanetary Phenomenon Unit to study the crashed remains of one of the Los

Angeles Air Raid craft, what did the Navy do with the craft it retrieved from the Pacific Ocean? According to information provided by Tompkins, the Navy created its own intelligence program to gather as much data about the flying saucer phenomenon as possible using its worldwide intelligence assets. This was especially important when it came to investigating intelligence reports of Nazi Germany developing flying saucer prototypes.

Figure 6. Hoover's Handwritten Comment on FBI Memo

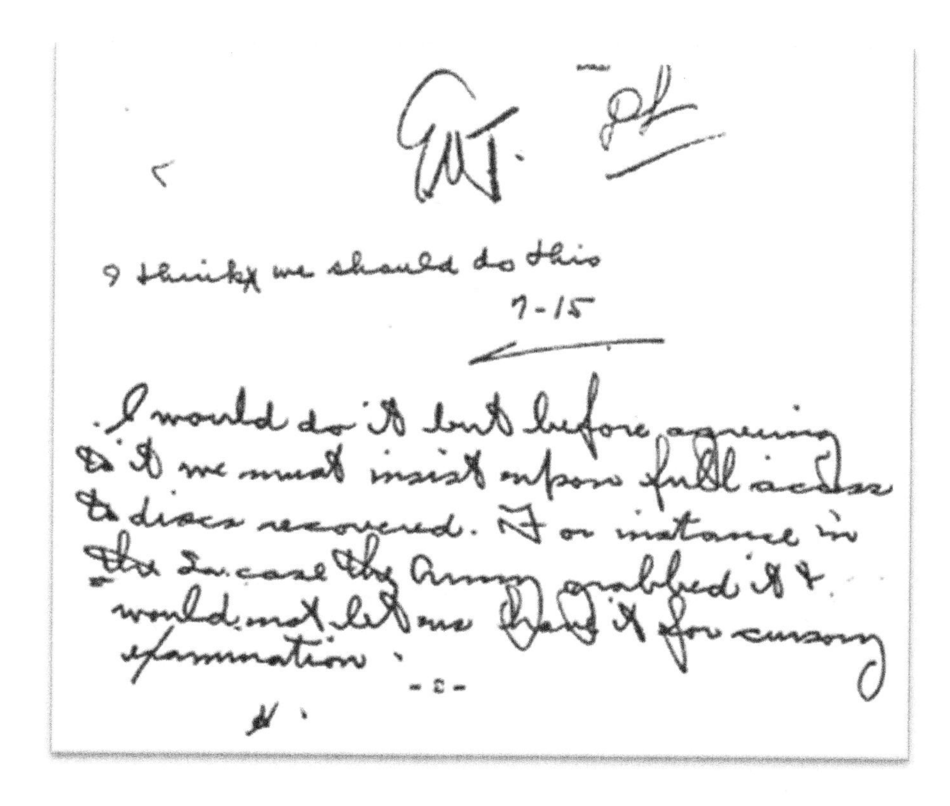

I think we should do this
7-15

I would do it but before agreeing to it we must insist upon full access to discs recovered. For instance in the La. case the Army grabbed it & would not let us have it for cursory examination.

- 2 -

ENDNOTES

[6] Wesley Frank Craven and James Lea Cate, eds., *The Army Air Forces in World War II* (Office of Air Force History, 1983) p. 283. Available online at: http://tinyurl.com/jcxxmu8
[7] Wesley Frank Craven and James Lea Cate, eds., *The Army Air Forces in World War II*, p. 284. Available online at: http://tinyurl.com/jcxxmu8
[8] Wesley Frank Craven and James Lea Cate, eds., *The Army Air Forces in World War II*, p. 284. Available online at: http://tinyurl.com/jcxxmu8
[9] Wesley Frank Craven and James Lea Cate, eds., *The Army Air Forces in World War II*, p. 285. Available online at: http://tinyurl.com/jcxxmu8
[10] William Tompkins, *Selected by Extraterrestrials: My life in the top secret world of UFOs, think-tanks and Nordic secretaries* (Createspace, 2015) p. xi.
[11] "The Great Los Angeles Air Raid," http://theairraid.com/
[12] See Tompkins, *Selected by Extraterrestrials*, p. xii.
[13] "The Great Los Angeles Air Raid," http://theairraid.com/
[14] William Tompkins covert work with the Office of Naval Intelligence will be discussed in chapter two.
[15] Phone interview with William Tompkins, September 19, 2016.
[16] Available online at: http://majesticdocuments.com/pdf/marshall-fdr-march1942.pdf
[17] Available online at: http://majesticdocuments.com/pdf/marshall-fdr-march1942.pdf
[18] Available online at: http://www.textfiles.com/ufo/UFOBBS/1000/1723.ufo
[19] Timothy Good, *Above Top Secret*, p. 484.
[20] Available online at: http://majesticdocuments.com/pdf/marshall-fdr-march1942.pdf
[21] Robert and Ryan Wood, http://majesticdocuments.com/documents/pre1948.php
[22] According to Timothy Good, prior to the documents release, the Department of Defense denied having any further information about the Los Angeles Air Raid, see *Above Top Secret: The World Wide U.F.O. Cover Up* (Quill, 1988) p. 17.
[23] Quoted in Timothy Good, *Above Top Secret*, p. 17.
[24] Document available online at: http://aboutfacts.net/ufo/UFO43/Small/HooverUFO.jpg

CHAPTER 2

COVERT NAVY ESPIONAGE PROGRAM ON NAZI GERMANY'S FLYING SAUCER PROJECTS

Mussolini's Flying Disk

With the formal entry of the US into War World II on December 7, 1941, the two main military services of the United States, the Army and Navy, would rapidly expand their global intelligence gathering operations to learn about the latest weapons and technology programs of the Axis Powers. Then came the extraordinary February 24/25, 1942 Los Angeles Air Raid UFO incident, which propelled the Army's G2 to set up the Interplanetary Phenomenon Unit to investigate reports of "flying saucer" technologies being seen or developed anywhere that the US military could gain access.

On a parallel course, the Office of Naval Intelligence would set up its own intelligence gathering operation to investigate the UFO phenomenon using its world-wide assets. Both the Army and Navy were clearly concerned over the possibility that the Axis Powers were developing technologies reflected in what had been seen and recovered during the Los Angeles Air Raid incident.

Yet, the first documented study of an advanced flying saucer craft can be traced to Fascist Italy almost a decade earlier. In the late 1990's, 18 Fascist-era documents were leaked to Italian researchers dating from the 1930's, which dealt with UFOs.[25] These files were allegedly official documents leaked by someone who had inherited them from a former top level insider in the Italian Intelligence community. Of paramount importance, the Italian documents were originals, which allowed forensic analysis to determine their age and authenticity. After forensic analysis was completed and dismissed the possibility that they were modern forgeries; Italian researchers released the documents and their findings.

Among the files were documents mentioning a 1933 UFO crash in the Lombardy region involving a highly advanced craft that resembled a flying saucer. A top secret organization called Cabinet RS/33 was created to study the captured craft, along with the growing number of flying saucer sightings.[26] It was headed by famed Italian inventor, Guglielmo Marconi.

As Cabinet RS/33 continued its classified study of the captured 1933 flying saucer and sightings of similar craft in the immediate years after, advisors at some point would present Mussolini with the determination that Nazi Germany was involved. The 1933 flying saucer in possession of the Italians, and/or later UFO sightings, were either part of or related in some way to a top secret Nazi advanced aviation program.

Once Mussolini's Cabinet RS/33 realized that it was Nazi Germany, and not France or Britain, behind the flying saucer phenomenon, this would surely influence Fascist Italy as it feverishly modernized its Air Force with the increasing likelihood of another major European war. Discovery of a flying saucer and related craft belonging to a secret space program created by Germany explains Mussolini's puzzling decision to increasingly ally himself with Hitler.

A secret agreement between Hitler and Mussolini for the study and development of flying saucer technologies was reached sometime in 1938, according to one of the leaked Italian UFO documents:

> ... in this latest consignment of material, he included copies of new documents which—so he said—demonstrated the existence of agreements between Hitler and Mussolini for the study of alien technology, agreements that had been made in 1938; these documents were: an Agency Stefani message from Florence containing an interview with the Fuhrer Hitler when he was visiting Italy; a banknote of the nominal value of a million Lire (maybe "black funds of the RS/33 CABINET"); minutes regarding the oath of secrecy given by the professors who collaborated with the Fascist Government; an invitation (registered) to Benito and Rachele Mussolini to Villa Torlonia (said by "MR. X" to be for ... an extremely private meeting dedicated to the RS/33 CABINET).[27]

The secret agreement between Fascist Italy and Nazi Germany, to cooperate in the development of flying saucers and a secret space program, would have come to the attention of the military intelligence services of major European nations such as Britain, France and the Soviet Union on the eve of World War II. In particular, Winston Churchill corresponded from 1936 to 1940 with Benito Mussolini in a failed attempt to dissuade him from establishing an alliance with Hitler. It's very likely that at some point Churchill learned something about what the Italians were doing in the field of advanced aeronautics. After World War II, Churchill, allegedly with the aid of Britain's MI6, traveled to Northern Italy to find his private correspondence with Mussolini and have it destroyed.[28]

As Britain and the United States increasingly cooperated after the start of armed hostilities on September 1, 1939, British intelligence would share its findings about the Nazi Fascist cooperation on flying saucer programs. A possible connection to the craft involved in the Los Angeles Air Raid incident required serious investigation by the US military intelligence services.

A covert espionage program was subsequently created by the Office of Naval Intelligence to infiltrate the advanced aerospace programs of Nazi Germany, which were known to be developing antigravity vehicles for possible use in World War II and even for space flight. Covert naval operatives were embedded inside a number of Nazi Germany's most advanced research and development programs, and tasked to report back what they had seen in debriefings secretly held at Naval Air Station, San Diego. This information was then relayed by an appointed "Information Disseminator" to various US think tanks, aircraft companies and university departments, where scientists and engineers attempted to understand the principles and the war potential of the Nazi antigravity technology projects. Now, we finally know all this occurred since in December 2015, the Navy's Information Disseminator publicly revealed himself.

William Tompkins: Information Disseminator for Navy's Covert Espionage Program out of Nazi Occupied Europe

William Tompkins has revealed that he worked as the "Information Disseminator" for the US Navy's covert espionage program out of Nazi Germany from 1942 to early 1946. To support his claim, he has supplied a mission statement dated September 26, 1945, which he explains was retroactively applied to his wartime service. Tompkins' mission statement is documentary support for his assertion that he was a participant in the covert Navy espionage program set up in Nazi Occupied Europe to learn about advanced aircraft programs, some of which he described as using antigravity principles.

Tompkins' story begins in 1932, when he was only nine. He recounts how his father took him to the Navy's deep water harbor in Long Beach, California where he began the first of many public tours of aircraft carriers, battleships, cruisers and destroyers, which were anchored. The public was not allowed to take photos of the ships whose gun placements and radar designs were still classified.

Figure 7. Tompkins' Mission Orders

U.S. NAVAL AIR STATION
SAN DIEGO 35, CALIFORNIA

26 September 1945.

STATEMENT OF MISSION, TASKS AND OBJECTIVES

DISSEMINATOR OF AIRCRAFT RESEARCH AND INFORMATION

Shop 160 - Planning Division

WILLIAM M. THOMPKINS, USNR - 680-52-78

MISSION:- Under the direction of the Production Superintendent. In addition to reporting directly to the Production Superintendent, the Disseminator of Aircraft Research and Information shall also report to the Planning Division Superintendent and to the Chief Engineer to coordinate, compile and maintain a continuous survey of research and information relative to special equipment necessary in the repair and overhaul, experimental tests, and developmental work of aircraft, air craft engines and their accessories.

Tompkins Mission Order as Disseminator of Aircraft Research & Information
Source: William Tompkins, *Selected by Extraterrestrials*, p. 314

Possessing a photographic memory, Tompkins tells how he walked all around the ships and took notes of what he observed, and soon after began reproducing scaled models. Eight years later, Tompkins had built many detailed and flawlessly accurate models of the Navy ships he had seen, and his father was subsequently displaying them in a Hollywood Department store.

In 1941, after the Navy learned about the model ships displaying top secret details, they interrogated Tompkins' father rigorously, and then Tompkins, himself. The models were quickly withdrawn from public circulation. Senior Navy officers were nevertheless greatly impressed by Tompkins' remarkable talent.

Figure 8. Tompkins' Ship Models on Display in Hollywood

A March 26 article in the *Evening Outlook* newspaper of Santa Monica featured a photograph of Tompkins (age 17 at the time) showing some of his ship models to Navy Captain G.C. Gearing, Commandant of the 11th Naval District in San Diego.

Rear Admiral C. A. Blakeley was quoted in the *Evening Outlook*:

> It is with considerable interest and pleasure that I, together with officers of my staff, examined several of the ship models. Craftsmanship such as you have evidenced shows that you are a keen student of detail and naval construction. Best of all, however, you are doing something worthwhile as a young American—you are helping to build into the American mind the importance of the nation's first line of defense to each American, young and old.[29]

Admiral Blakely had written directly to Tompkins on March 10, 1941, directly congratulating him on his accomplishment.

Figure 9. Bill Tompkins Showing his Models to Captain Gearing

EVENING OUTLOOK, SANTA MONICA, CALIFORNIA

Youth Models Ships Of American Fleet

Naval Officers Praise His Work

Carving of units of the U. S. fleet from balsa wood, started three years ago by 17-year-old William M. Tompkins, has created a furor among naval officers.

Young Tompkins, who resided at 833 21st st. for six years and attended school in Santa Monica, now lives at 3224 Ellington Drive, Hollywood. The family moved from Santa Monica three years ago, but there still are relatives here.

HAS 51 SHIPS

In all he has 51 ships, each made on a scale of 1 inch to 50 feet and ranging from 18½ inches for the aircraft carrier Lexington to 1½ inches for shore boats. The Lexington iof this miniature fleet carries 28 perfectly modeled fighting planes on its deck.

Tompkins exhibited his fleet to navy officers in San Diego. Capt. H. C. Gearing, commandant of the 11th Naval District, was so impressed by the fidelity to detail that he arranged to borrow them for display at the Naval Training Station in San Diego.

The "fleet" is made up of four battleships, the New York, Idaho, Oklahoma and West Virginia; the airplane carrier, 11 destroyers and many light and heavy cruisers, submarines, tenders, repair ships and other units that go to make up the country's first line of defense.

MINIATURE FIGHTING SHIPS modeled by 17 year-old William M. Tomkins, formerly of Santa Monica, have aroused the interest and admiration of naval officers. Photo shows the youth displaying his fleet to Capt. H. C. Gearing, commandant of the 11th Naval District San Diego.

Tompkins' photographic memory and ability to reproduce complex ship design was extraordinary, and a US Naval Intelligence Officer, Lt. Perry Wood, understood the contribution Tompkins could make to the Navy's pursuit of advanced technology:

> Early in 1942, naval intelligence officer Lt. Perry Wood, understanding the technical capabilities and historic research necessary to create the ship models, put together a mission package that resulted in Bill's induction into the navy. After completing boot camp in San Diego he was assigned a position in naval intelligence on advanced technology projects.[30]

Fatefully, Tompkins' abilities would come to the attention of senior officials looking for the right person to participate in the debriefings of spies involved in a covert espionage program out of Nazi Germany and its occupied territories. Tompkins says that Rear Admiral Rico Botta placed him on his personal staff for four years (1942-46), and quickly elevated him to a leadership position in the covert intelligence program, which used Naval operatives with the rank of Lieutenant. All were second generation Americans with German ancestry, which allowed them to easily infiltrate Nazi Germany.

Though officially only a seaman (E-3) when he began the covert operation, Tompkins says he replaced a Commander at North Island Naval Air Station in San Diego, who previously performed the role of "Disseminator of Aircraft Research and Information" in the espionage program. According to Tompkins, the spies, which numbered 29, travelled in to San Diego on a six monthly rotational basis to give debriefings to a small group comprising the head of the program, Admiral Botta, three captains, a stenographer, and himself. Tompkins described how the meetings occurred:

Figure 10. Admiral Blakely's Letter to Bill Tompkins

REPLY KINDLY ADDRESS
"COMMANDANT"
THE SIGNER BY NAME

COMMANDANT'S OFFICE
ELEVENTH NAVAL DISTRICT
SAN DIEGO, CALIFORNIA

March 10, 1941

Dear Mr. Tompkins:

It is with considerable interest and pleasure that I, together with officers of my staff, examined several of the ship models of your miniature United States Fleet.

Craftsmanship such as you have evidenced in building "your navy" shows that you are a keen student of detail and naval construction. Best of all, however, you are doing something worthwhile as a young American---you are helping to build into the American mind the importance of the Nation's First Line of Defense to each American, young and old.

We of the Navy enjoyed today's visit from "your fleet" for we recognized so much of your true detail copied from ships we have served in.

With every good wish for a career of success and happiness, I am

Very sincerely,

Chas A. Blakely
C. A. BLAKELY
Rear Admiral, U.S.N.
Commandant, 11th Naval District

Mr. William M. Tompkins
3224 Ellington Drive
Los Angeles, California

> We had this little small office. He [Rico Botta] was commander of Naval Air Station, San Diego and so the admiral would be sitting here. I'm sitting next to him and one of my three captain bosses sitting here. A typist was sitting at the other side of the table and the Navy operative — a Lieutenant Commander or Lieutenant would sit over there. And he would lay out the information [with] some sketches, very few photographs, so it was almost all verbal, almost all. And if it was a system that we had already known about, other operatives from the information two years before, and it was an update of it, he would tie together all of the other organizations that were tied with that specific part, of say the cigar shaped vehicles, what the weapons were, what the operations were, the different companies underneath the ground, which were slave labor organizations.[31]

With the utmost care and attention, Tompkins would listen to the spies' debriefings and subsequently create a detailed briefing packet, which he would personally carry to select US aerospace facilities using a private aircraft supplied to him by Admiral Botta. Tompkins has explained more of this process in an interview:

> ... a Navy guy comes back from Germany, he tells us about this, we put those packages together. I had 10 or 12 girls, typists, put together the packages. The package is not to do it, the package is to take it to Caltech and "you study it." Take the package to Lockheed, "what can you do with this?"[32]

He has also identified some of the Navy research facilities that the packages were taken to:

> We took the data to every Navy Research organization. We took it to Navy Development Center, Warminster,

Pennsylvania. OK. That's a big facility. I bet I went 20 times back there with stuff. The Caltech one, I didn't even go to Pasadena for that. I went to the Navy's weapons system at China Lake because Caltech was working on other programs there. They would pull their scientists off, whatever that was, and throw these guys into a package, and run with the ball...[33]

Figure 11. Tompkins' with Four of his Typist Assistants

1943 photo of Bill Tompkins with typists that helped prepare Briefing Packets for dissemination. His magnified badge shows details of Navy unit used for the covert operation

In support of his claim, Tompkins has supplied copies of two separate passes he received to enter and leave the San Diego Naval Air Station with up to three packages. These packages contained the alleged secret data provided by the Navy agents being distributed by Tompkins to select think tanks and corporations. The two passes in the following image show Tompkins was permitted to carry one and three (briefing) packets out of the Naval Air Station, respectively on April 17, 1944 and December 24, 1943.

Figure 12. Tompkins' Two Exit Passes from Naval Air Station, San Diego

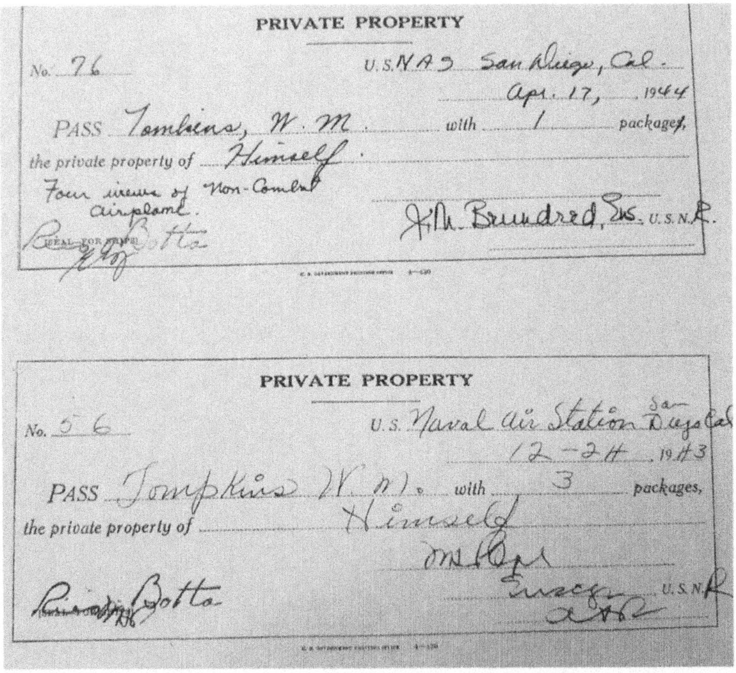

PRIVATE PROPERTY

No. 76 U.S. NAS San Diego, Cal.

Apr. 17, 1944

PASS Tomkins, W. M. with 1 packages,

the private property of Himself.

Four views of Non-Combat airplane.

J.M. Brundred, Ens. U.S.N.R.

PRIVATE PROPERTY

No. 56 U.S. Naval Air Station San Diego Cal

12-24 1943

PASS Tompkins W. M. with 3 packages,

the private property of Himself

Ensign U.S.N.R.

Passes for travel from Naval Air Station with briefing packages
Source: William Tompkins, *Selected by Extraterrestrials*, p. 68

In addition to the signature of the final authorizing officer, Admiral Rico Botta (bottom left) in itself is highly significant; close examination of the April 1943 exit pass provides additional crucial corroborating information for Tompkins' story. In his book, Tompkins says that he was given access to the smaller of two planes that were at the disposal of Admiral Botta. Tompkins then explains that he used the plane to take the briefing packages to different aerospace facilities where the packages were used for research of extraterrestrial related technologies:

> Flew admiral's plane to Douglas El Segundo, Lockheed Burbank and China Lake for four years... 1943-1945. Spent much time at China Lake facility modifying rockets for testing (40 visits) and working on their alien projects.[34]

The exit pass (see Figure 13) says "four [for] issue of non-combat aircraft." This corroborates his claim that he was authorized to expeditiously take the briefing packages to a select number of aviation companies, think tanks and university departments. The April 1944 exit pass authorizing Tompkins to carry a packet using a non-combat aircraft is highly significant in terms of demonstrating that he performed covert assignments as a "Disseminator of Aircraft Research and Information," in addition to his official duties at the Naval Air Station.

The next document (see Figure 14), issued just prior to his honorable discharge in 1946, shows Tompkins completing a final work assignment during World War II as an Aviation Machinist's Mate Petty Officer 2nd Class (AMM2).

According to the Bureau of Naval Personnel Bulletin of May 1944, the official duties of an Aviation Machinist's Mate were: "Maintains and repairs aircraft engines, propellers, fuel systems, brakes, hydraulic system, gears, starters. Operates machine-shop tools."[35] Carrying "packets" to unknown locations, and using non-combat aircraft to do so, is clearly well beyond the responsibilities of an Aviation Machinist's Mate.

Figure 13. Analysis of April 17, 1944 Exit Pass

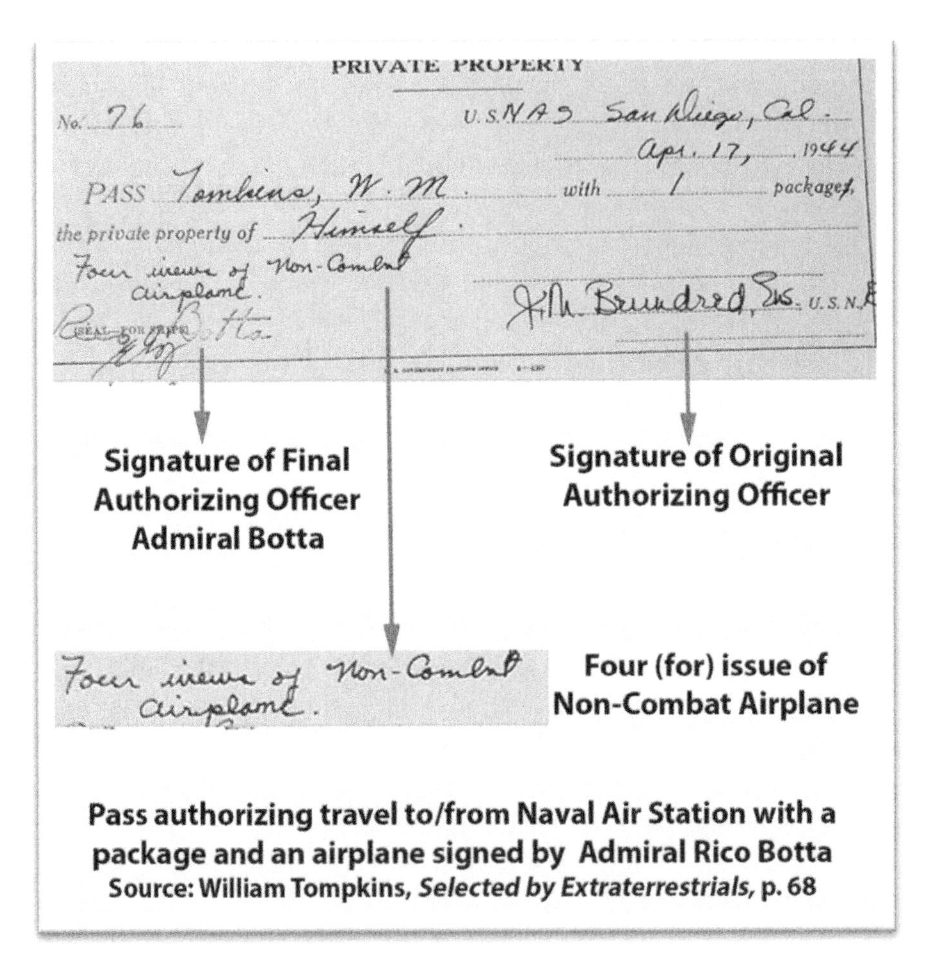

Pass authorizing travel to/from Naval Air Station with a package and an airplane signed by Admiral Rico Botta
Source: William Tompkins, *Selected by Extraterrestrials*, p. 68

Figure 14. Tompkin's Final Duty Before Honorable Discharge

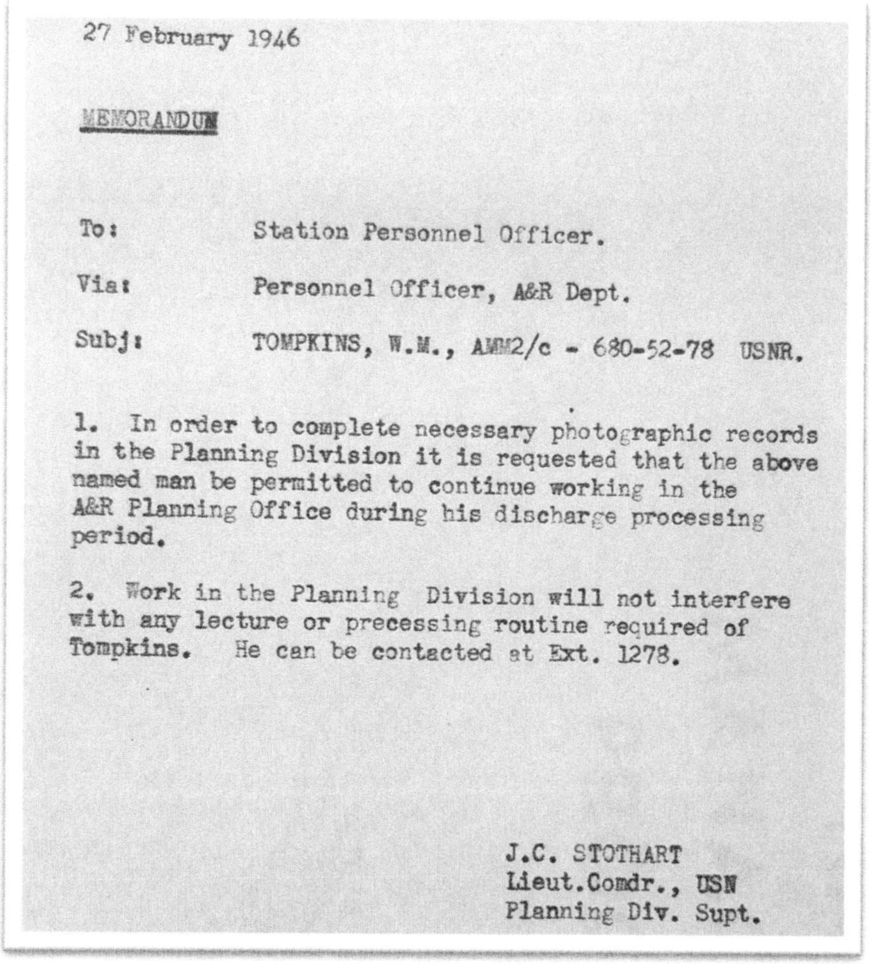

27 February 1946

MEMORANDUM

To: Station Personnel Officer.

Via: Personnel Officer, A&R Dept.

Subj: TOMPKINS, W.M., AMM2/c - 680-52-78 USNR.

1. In order to complete necessary photographic records in the Planning Division it is requested that the above named man be permitted to continue working in the A&R Planning Office during his discharge processing period.

2. Work in the Planning Division will not interfere with any lecture or precessing routine required of Tompkins. He can be contacted at Ext. 1278.

J.C. STOTHART
Lieut.Comdr., USN
Planning Div. Supt.

Furthermore, Tompkins' two exit passes display a signature which he states belonged to the commander of the Assembly and Repair Department, Naval Air Station, San Diego—Admiral Rico Botta. If it was indeed Botta that signed the two exit passes, this would help considerably in confirming Tompkins' claims of performing covert missions in addition to his official duties. Botta's signature appears on a number of documents received from the National Archives under the Freedom of Information Act.[36] One document shows a picture of Botta dating from 1934 when he was a Lieutenant Commander.

Figure 15. Botta's Signature on 1934 Document

Rico Botta, Lieutenant Commander, USN
27 August, 1934.
Rico Botta
Signature.

The 1934 document contains a clear image of his signature which can be compared to the signature on the two exit passes supplied by Tompkins. As the following graphic shows, and as document verification expert Dr. Robert Wood confirms, the signatures look identical.[37]

The confirmation that it is indeed Botta's signature which appears on the exit passes is documentary evidence supporting Tompkins claim that he was performing duties as a disseminator of "Naval Aircraft Research and Development" as described in his mission orders. In addition, Tompkins was allowed to fly an airplane in support of his mission.

Figure 16. Comparison of Botta's Signatures

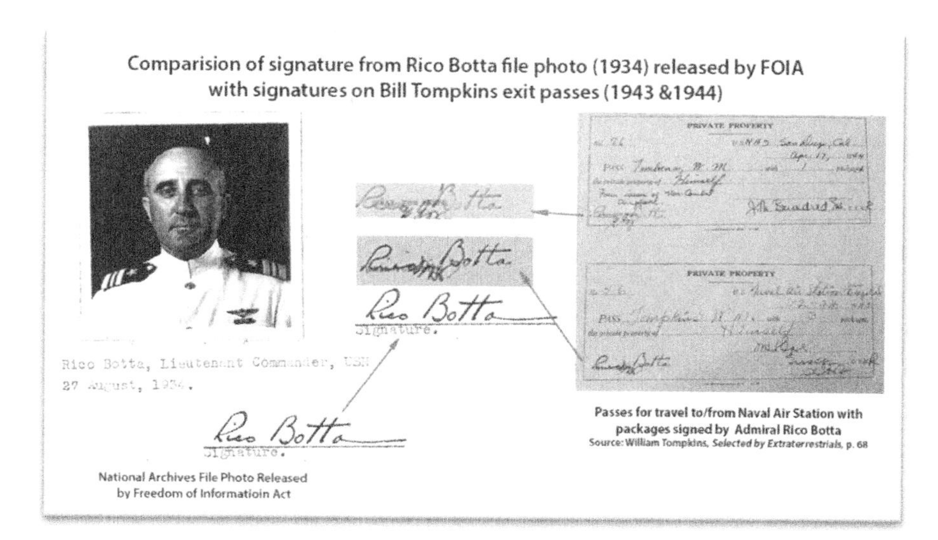

Botta is a crucial figure whose exceptional job performance was recognized in running the San Diego Naval Air Station and spy program. He was in charge of five thousand personnel performing maintenance, repair and upgrades for the Navy's top combat aircraft at the start of World War II. By the end of the war, the Assembly and Repair Department had expanded to over ten thousand personnel. The Admiral's exemplary technical knowledge and leadership at the vital San Diego base led him to be awarded a Legion of Merit in 1945 (see Figure 17).

It is no small fact to be missed that the distinguished commander, with such vast responsibilities, is the one who signed Tompkins' exit passes. A lesser ranked superior would normally sign off on such minor documentation. Clearly Tompkins was operating within a position of key importance to the Admiral, and it placed him directly under Botta's wing within the covert intelligence program.

Figure 17. Legion of Merit Awarded to Rico Botta

IN REPLY REFER TO NO.
NA11/OO/P15 LEG:gvw
Serial: 29-10

U.S. NAVAL AIR STATION
SAN DIEGO 35, CALIFORNIA

To: Secretary of the Navy.
Via: (1) Commander, Naval Air Bases, Eleventh Naval District.
(2) Commandant, Eleventh Naval District.
(3) Chief of Bureau of Aeronautics.

Subj: CAPTAIN RICO BOTTA, U.S.N. - Recommendation for Award of Legion of Merit.

Ref: (a) AlNav 291, 27 September 1945.

1. Captain Rico Botta, U.S.N. has served as the Assembly and Repair Officer for the Naval Air Station, San Diego, since 31 December 1942. At the time he assumed this duty the Assembly and Repair Department had 3,440 civilian employees and 1,925 enlisted men, a total of 5,365 personnel. At the termination of hostilities on 15 August 1945, this Department had expanded to 5,342 civilian employees and 5,219 enlisted personnel, a total of 10,561 personnel. During the year 1943, 520 naval aircraft were reconditioned in the A&R Department in addition to 6,426 aircraft on which modifications and minor repairs were made, including preparation for shipping to combat areas. In 1944, 989 aircraft were reconditioned in addition to 10,769 on which modifications and minor repairs were made. In 1945, up to 11 November, 1,243 aircraft were reconditioned in addition to 5,664 on which modifications and minor repairs were made, up to 1 September 1945. These figures indicate the tremendous expansion not only in the number of persons employed in the A&R Department, but in the numbers of aircraft worked over and made ready in this Department for combat service. In addition to this tremendous load of aircraft work, this Department was frequently called upon to accomplish many other emergency projects, such as building and constructing special bomb and ammunition storage racks in ships alongside the Air Station docks. One of the outstanding special rush projects which was undertaken and pushed to completion was the manufacture of a large quantity of a newly developed special device for aerial torpedoes which resulted in the well-known and outstanding improvement in performance of aerial torpedoes launched against the Japanese fleet, and contributed in such a large degree to the defeat of the enemy in the battles of the Philippine Seas.

2. This Station, as the principal reconditioning and modification center for new carrier aircraft was called upon throughout the war to incorporate the latest changes dictated by combat experience in new aircraft assembled here for shipment to the forward areas. This invariably required the development of new tools and new processes, always working against a strict time limit, in order that there would be no delay in shipment of sorely needed combatant aircraft to the forward areas. Month after month, the Assembly and Repair Department of this Station, under Captain Botta's outstanding leadership, and as a result of his superior technical knowledge

- 1 -

Admiral Rico Botta's Special Expertise & Secret Intelligence Role

It is critical to connect Botta to the retrieval of craft from the Los Angeles Air Raid incident, and uncover the reason why he was chosen to head the debriefings of covert Naval personal reporting on secret Nazi activities. The next FOIA document (see Figure 18) reveals that on February 25, 1942, the day of the Los Angeles Air Raid, Botta was ordered to travel from the Navy's Bureau of Aeronautics, in Washington DC, to Wright Field, Dayton, Ohio. At the time, Wright Field was the location for both the Army Air Force and Navy's study of foreign and experimental aircraft. It was the place where either or both Army G2 and the Office of Naval Intelligence would have transferred the craft retrieved from the night of February 24/25.

The experimental aircraft facilities at Wright Field was the very same location that the craft, later retrieved from the July 1947 Roswell UFO incident, would be taken according to multiple documents and whistleblowers. An official FBI teletype, dated July 8, 1947 (see Figure 19), confirms that UFO debris from the Roswell incident was indeed immediately taken to Wright Field after its discovery, despite the Army Air Force's misleading claim that it was a mistaken weather balloon:

> EIGHT AIR FORCE, TELEPHONICALLY ADVISED THIS OFFCE THAN AN OBJECT PURPORTING TO BE A FLYING DISC WAS RECOVERED NEAR ROSWEL, NEW MEXICO.... DISC AND BALLOON BEING TRANSPORTED TO WRIGHT FIELD BY SPECIAL PLANE FOR EXAMINATION. INFORMATION PROFIDED THIS OFFICE BECAUSE OF NATIONAL INTREST IN CASE.[38]

Figure 18. Document Showing Botta's Travel to Wright Field Authorized on Feb 25, 1942

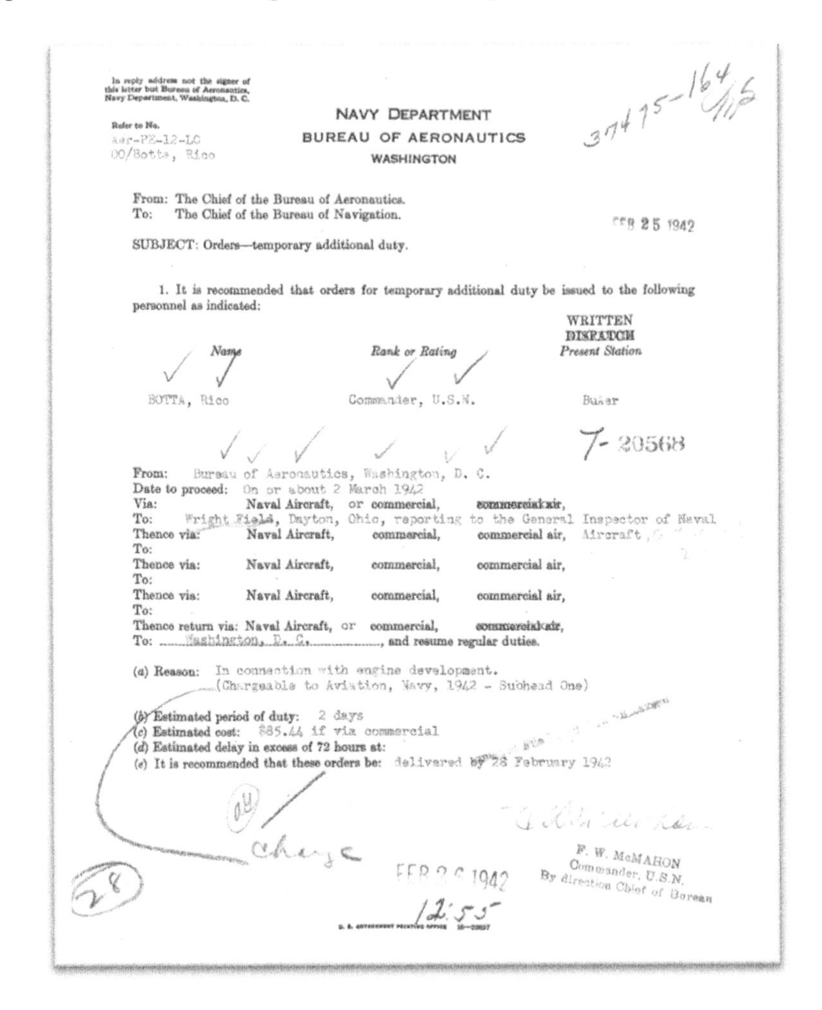

In reply address not the signer of this letter but Bureau of Aeronautics, Navy Department, Washington, D. C.

Refer to No.
Aer-PE-12-LC
00/Botta, Rico

NAVY DEPARTMENT
BUREAU OF AERONAUTICS
WASHINGTON

From: The Chief of the Bureau of Aeronautics.
To: The Chief of the Bureau of Navigation.

FEB 25 1942

SUBJECT: Orders—temporary additional duty.

1. It is recommended that orders for temporary additional duty be issued to the following personnel as indicated:

Name	*Rank or Rating*	WRITTEN DISPATCH *Present Station*
BOTTA, Rico	Commander, U.S.N.	BuAer

7-20568

From: Bureau of Aeronautics, Washington, D. C.
Date to proceed: On or about 2 March 1942
Via: Naval Aircraft, or commercial, ~~commercial air~~,
To: Wright Field, Dayton, Ohio, reporting to the General Inspector of Naval Aircraft,
Thence via: Naval Aircraft, commercial, commercial air,
To:
Thence via: Naval Aircraft, commercial, commercial air,
To:
Thence via: Naval Aircraft, commercial, commercial air,
To:
Thence return via: Naval Aircraft, or commercial, ~~commercial air~~,
To: Washington, D. C., and resume regular duties.

(a) Reason: In connection with engine development.
(Chargeable to Aviation, Navy, 1942 - Subhead One)

(b) Estimated period of duty: 2 days
(c) Estimated cost: $85.44 if via commercial
(d) Estimated delay in excess of 72 hours at:
(e) It is recommended that these orders be: delivered by 28 February 1942

F. W. McMAHON
Commander, U.S.N.
By direction Chief of Bureau

FEB 26 1942

U. S. GOVERNMENT PRINTING OFFICE 16—20607

Figure 19. FBI Teletype Shows Roswell UFO Debris Was Taken to Wright Field

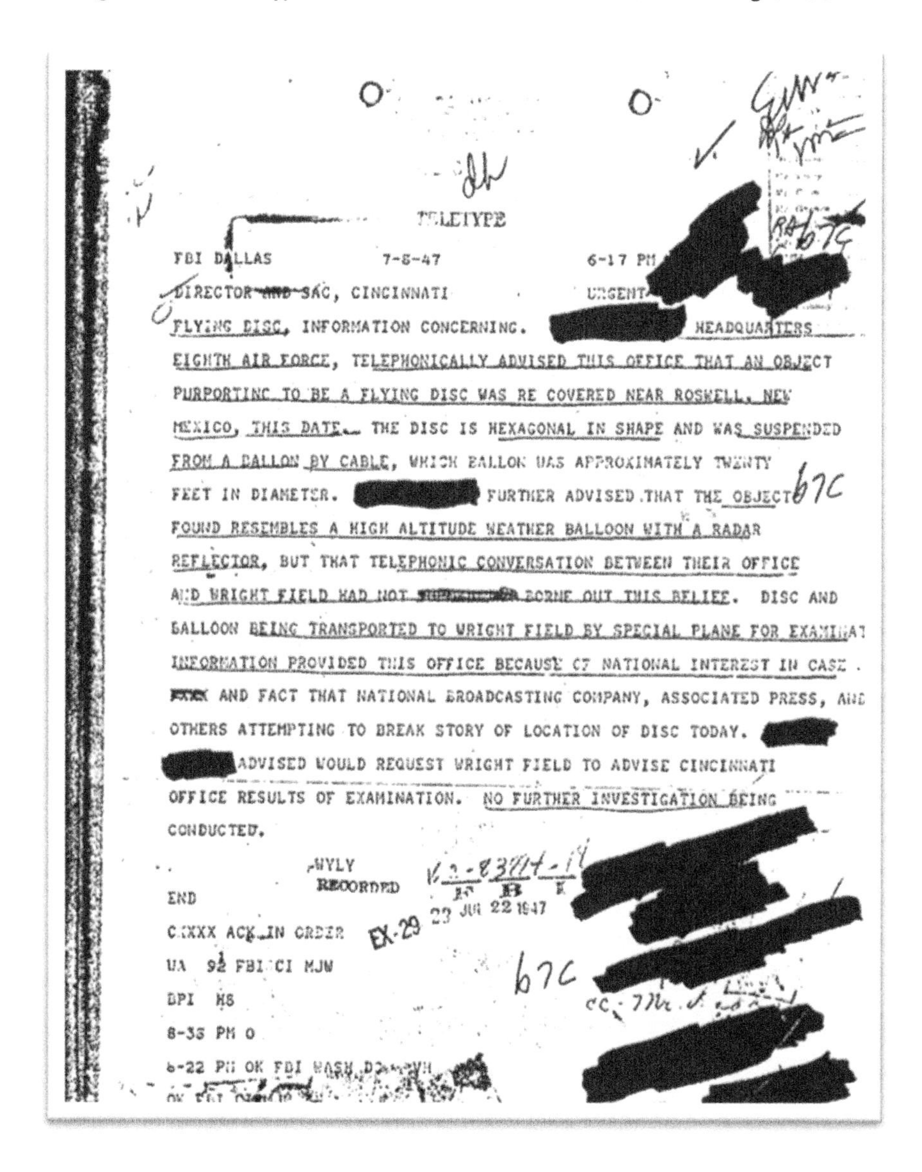

TELETYPE

FBI DALLAS 7-8-47 6-17 PM

DIRECTOR AND SAC, CINCINNATI URGENT

FLYING DISC, INFORMATION CONCERNING. [redacted] HEADQUARTERS EIGHTH AIR FORCE, TELEPHONICALLY ADVISED THIS OFFICE THAT AN OBJECT PURPORTING TO BE A FLYING DISC WAS RE COVERED NEAR ROSWELL, NEW MEXICO, THIS DATE. THE DISC IS HEXAGONAL IN SHAPE AND WAS SUSPENDED FROM A BALLON BY CABLE, WHICH BALLON WAS APPROXIMATELY TWENTY FEET IN DIAMETER. [redacted] FURTHER ADVISED THAT THE OBJECT FOUND RESEMBLES A HIGH ALTITUDE WEATHER BALLOON WITH A RADAR REFLECTOR, BUT THAT TELEPHONIC CONVERSATION BETWEEN THEIR OFFICE AND WRIGHT FIELD HAD NOT BORNE OUT THIS BELIEF. DISC AND BALLOON BEING TRANSPORTED TO WRIGHT FIELD BY SPECIAL PLANE FOR EXAMINAT INFORMATION PROVIDED THIS OFFICE BECAUSE OF NATIONAL INTEREST IN CASE AND FACT THAT NATIONAL BROADCASTING COMPANY, ASSOCIATED PRESS, AND OTHERS ATTEMPTING TO BREAK STORY OF LOCATION OF DISC TODAY. [redacted] ADVISED WOULD REQUEST WRIGHT FIELD TO ADVISE CINCINNATI OFFICE RESULTS OF EXAMINATION. NO FURTHER INVESTIGATION BEING CONDUCTED.

WYLY

RECORDED

END

23 JUL 22 1947

CXXXX ACK IN ORDER

EX-29

UA 92 FBI CI MJW

b7C

DPI HS

8-35 PM O

6-22 PM OK FBI WASH

UFO researchers, Tom Carey and Donald Schmitt, wrote the book, *Inside the Real Area 51: The Secret History of Wright Patterson*, detailing the substantial evidence linking Wright Field, and its successor, Wright Patterson Air Force Base, as the premier US Air Force installation for studying recovered antigravity craft. They wrote:

> Outside of the UFO community, few are aware of Wright-Patterson's distinction of maintaining the U.S. government's official investigation of the UFO phenomenon from 1947 until 1969. With its vital experience of testing and reverse-engineering all materials both foreign and "from space," it is a historic fact that whatever crashed outside of Roswell fell under their purview. It is also a documented fact that the "debris" from New Mexico was sent to Wright Field...[39]

The February 25, 1942 document confirms that on the day the Army Air Force captured and began the process of shipping the captured Los Angeles flying saucer craft to Wright Field, Botta received his orders to travel there. He arrived on March 2, and stayed for at least two days examining the captured Los Angeles UFO craft on behalf of the Navy's Bureau of Aeronautics.

The significance of the document cannot be underestimated since it reveals that Botta was identified by the Navy's leading engineering division, the Bureau of Aeronautics, as their representative for investigating the power systems of retrieved antigravity craft. At the time, Botta, then a Commander, headed the Bureau's Power Plant Design Branch. His technical expertise and knowledge of massive power systems was well known to his Bureau superiors. After the War, he received a Citation for his exemplary service as the head of this Branch, as the following February 18, 1946 document shows.

Figure 20. Botta's Citation for Exemplary Work in Navy Power Plant Design Branch

Prepared--7 Feb 1946

Pers 328--wcf Signed: 8 FEB 1946

The Secretary of the Navy takes pleasure in commending

CAPTAIN RICO BOTTA
UNITED STATES NAVY

for service as set forth in the following

CITATION:

"For outstanding performance of duty as Head of the Power Plant Design Branch in the Engineering Division, Bureau of Aeronautics, prior to the outbreak of hostilities and until December 1942. Charged with the responsibility for the design and development of power plants of maximum effectiveness for Fleet aircraft, Captain Botta performed his exacting duties with unusual competence, technical skill and aggressive leadership. It was largely through his keen foresight and intelligent direction of the aircraft engines program that Naval Aviation entered the war with the best aircraft engines in the world. The improvements contributed directly by him were major factors in the extraordinary advancement of aircraft engines during the war. His professional skill, penetrating insight and steadfast devotion to duty acted as a stimulus to those with whom he served and reflect the highest credit upon Captain Botta and the United States Naval Service."

A copy of this citation has been made a part of Captain Botta's official record, and he is authorized hereby to wear the Commendation Ribbon.

Prepared by: Capt H.G. Patrick,
U.S. Navy, Retired,
Navy Dept Bd of Dec & Medals

James Forrestal
Secretary of the Navy

Copy to: Pers 328
Secretary's Files

Ref: BdAwdsMtg 27 Dec 1945
End-2 on BuAer ltr Aer-E-4-SBS
00/Botta,Rico., Ser 296464,
dtd 30 Oct 1945
SHORE BASED A-62

SELECTION BOARD

Finished File Pers-328

Yet, another FOIA document (see Figure 21) shows that on October 1, 1942, one month before he was assigned to Naval Air Station, San Diego, the now Captain Botta, still stationed at the Bureau of Aeronautics, visited five top West Coast research facilities working on experimental aircraft. This FOIA document, issued on September 30, 1942, confirms that Botta was directly involved in the study of experimental aircraft, and that this was connected to his impending reassignment to one of these facilities, Naval Air Station, San Diego, in December 1942.

Of special significance is the reference to the Douglas Aircraft Company, which according to Tompkins was heavily involved with the Navy and Army Air Force in studying the feasibility of antigravity research and development. He says this cooperation led to an informal working group comprising: three Douglas engineers, two generals and two admirals to create a plan for what to do with the UFO craft retrieved from the 1942 Los Angeles Air Raid. The working group culminated in the launch of Project RAND out of the Douglas Headquarters in Santa Monica, California, in October 1945. I will examine Tompkins' claims regarding Project RAND in the next chapter.

Finally, the above travel authorization document demonstrates that Botta had the background and experience to direct Tompkins to different experimental aircraft facilities to deliver briefing packets in a covert intelligence program which he directed. Regarding the reference to Botta's rank as Captain in the September 30, 1942 FOIA document (see Figure 22), US Senate records available online confirm that in 1946, Botta had been retroactively promoted to Rear Admiral as of June 1943, based on his war time responsibilities.[40] FOIA documents provide additional information, such as the following document signed by Botta where he accepted his retroactive promotion.

This confirms that at least seven months after he began his new assignment running the Repair and Maintenance facility at the

Figure 21. Experimental Aircraft Facilities Visited by Rico Botta in Oct 1942

In reply address not the signer of this letter, but Bureau of Navigation, Navy Department, Washington, D. C.
Refer to No.
Pers-313-JLK

NAVY DEPARTMENT
BUREAU OF NAVIGATION
WASHINGTON, D. C.
(NOW BUREAU OF NAVAL PERSONNEL)

September 30, 1942.

T-12455

From: The Chief of Naval Personnel.
To: Captain
Rico Botta, U.S.N.,
Bureau of Aeronautics,
Navy Department.

Via: The Chief of the Bureau of Aeronautics.

Subject: Temporary additional duty.

1. Proceed to the place (or places--in the order given) indicated below, for temporary duty. This is in addition to your present duties and upon the completion thereof you will return to your station:

On or about October 1, 1942, to Los Angeles, Calif., via commercial aircraft, and thence to the following places and such other places as may be necessary, in connection with inspecting experimental aircraft and for conferences in connection with aircraft matters:

U.S. Army Air Force Experimental Station, Muroc, Calif.
Douglas Aircraft Company, Inc., El Segundo, Calif.
Northrop Aircraft, Inc., Los Angeles, Calif.
Naval Air Station, San Diego, Calif.
Consolidated Aircraft Corporation, San Diego, Calif.

2. You are authorized to omit or revisit any of the above mentioned places or vary the above itinerary as may be deemed necessary.

3. A per diem of $6.00 in lieu of subsistence will be allowed while in an air travel status during your absence from your station.

Copy to:
Bu. Aero.

Detail Office.

RANDALL JACOBS

List of experimental aircraft facilities Rico Botta traveled to before his transfer to Naval Air Station, San Diego

Naval Air Station in late December 1942, Botta was given a temporary wartime promotion to Rear Admiral. Due to the contingencies of the war which rapidly expanded the number of flag officer positions, the normal official promotion process was suspended in many cases. Botta is among a list of 58 Navy Flag Officers who had to wait until after the completion of the war for the official process to resume, and receive retroactive commissions reflecting their wartime ranks.[41] FOIA documents further confirm

Figure 22. Rico Botta's Acceptance of Retroactive Promotion to Rear Admiral

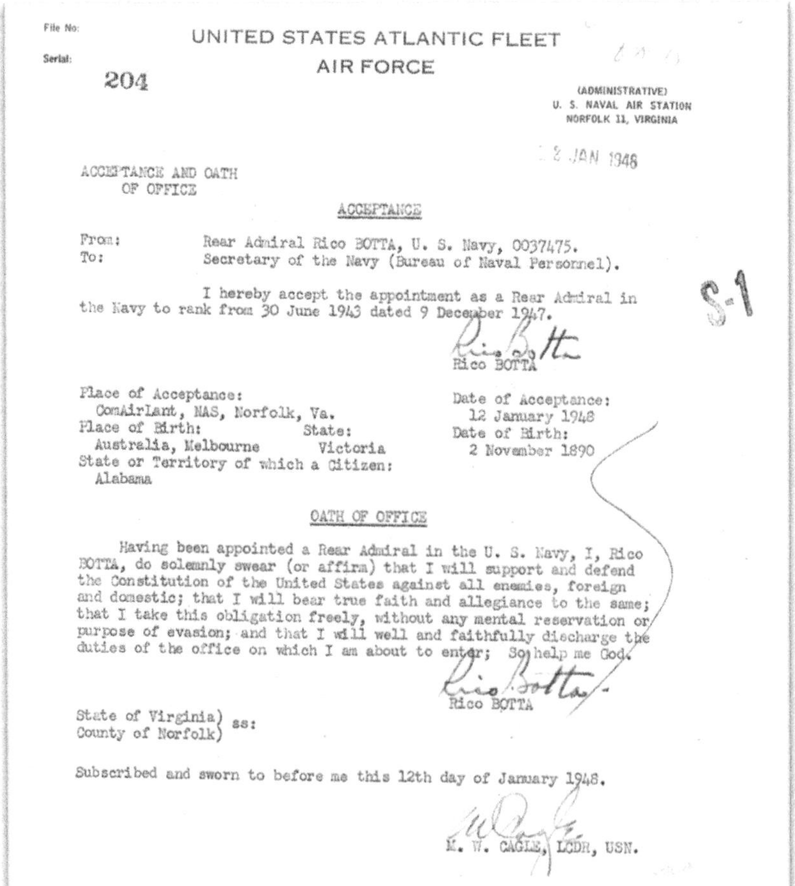

File No:
Serial: 204

UNITED STATES ATLANTIC FLEET
AIR FORCE

(ADMINISTRATIVE)
U. S. NAVAL AIR STATION
NORFOLK 11, VIRGINIA

12 JAN 1948

ACCEPTANCE AND OATH
OF OFFICE

ACCEPTANCE

From: Rear Admiral Rico BOTTA, U. S. Navy, 0037475.
To: Secretary of the Navy (Bureau of Naval Personnel).

I hereby accept the appointment as a Rear Admiral in the Navy to rank from 30 June 1943 dated 9 December 1947.

S-1

Rico BOTTA

Place of Acceptance:
ComAirLant, NAS, Norfolk, Va.
Place of Birth: State:
Australia, Melbourne Victoria
State or Territory of which a Citizen:
Alabama

Date of Acceptance:
12 January 1948
Date of Birth:
2 November 1890

OATH OF OFFICE

Having been appointed a Rear Admiral in the U. S. Navy, I, Rico BOTTA, do solemnly swear (or affirm) that I will support and defend the Constitution of the United States against all enemies, foreign and domestic; that I will bear true faith and allegiance to the same; that I take this obligation freely, without any mental reservation or purpose of evasion; and that I will well and faithfully discharge the duties of the office on which I am about to enter; So help me God.

Rico BOTTA

State of Virginia) ss:
County of Norfolk)

Subscribed and sworn to before me this 12th day of January 1948.

M. W. CAGLE, LCDR, USN.

that Botta retired in December 1952 with the rank of Rear Admiral Upper Class (a two star admiral) after completing his final assignment as the head of the Naval Air Material Center in Philadelphia.

Analysis of the previously shown documents corroborate Tompkins' claim that he was indeed acting as a courier ("Disseminator for Aircraft Research and Information") for a covert Naval Intelligence program with up to 29 spies in German occupied Europe during World War II. This leads to the question: What did Tompkins learn and what information was he carrying in the briefing packets to top US research and development facilities studying experimental aircraft?

Nazi Germany's Antigravity Programs & Extraterrestrial Alliances

What the naval operatives were reporting during their nighttime debriefings was completely unexpected, even astounding according to Tompkins. Admiral Botta and the three Navy captains could hardly believe what they were hearing. Tompkins says the operatives discovered that up to and during the World War II, there were two independent flying saucer programs under development in Nazi Germany. The first was largely a civilian effort that predated the Nazi rise to power in 1933, while the second was led by the Nazi SS.

Tompkins said that the civilian German space program had been inspired by a Nordic group of extraterrestrials, who were communicating through young female German mediums. In an interview, Tompkins confirmed that the leader of this group was Maria Orsic.[42] A stunning physical likeness between her and one of the Nordic aliens Tompkins says he met while working at Douglas Aircraft Company will be discussed in chapter four. Similarly, Corey

Goode says that when a photo of Maria Orsic was allegedly shown to the Swiss contactee, Billy Meier, he identified her as "Semjase," the Pleadian/Plejaran/Nordic extraterrestrial who was meeting with him. In a February 19, 2016 lecture, Goode said:

> ... when the military found out about Meier's case, they sent people over with some photographs for him to try and identify the female being he saw. He quickly pointed out one photograph, saying, "That's her! That's her!" Apparently the photo he pointed out was of Maria Orsic, the medium from the Vril Society, who was making contact with inner-Earth groups, and who played an intimate role in the pre and post World War II German secret space program.[43]

Orsic was allegedly sharing ideas and theories with leading inventors such as Nikola Tesla and Guglielmo Marconi. As mentioned earlier, Marconi headed the Italian Flying Saucer study group formed by Mussolini in 1933. Due to his association with Orsic, Marconi became aware of how advanced the civilian run German Secret Space program had become. After passing on some of this information to Mussolini, who was likely influenced by it to form an alliance with Hitler, Marconi decided to fake his death and disappear to South America to set up a civilian flying saucer program.[44]

Tompkins went on to explain that there were many inspired by Orsic's information, and/or the alleged Aldebaran extraterrestrials she was working with in the 1920's and 1930's. The goal was to build vehicles capable of deep space flight using principles of teleportation, and to even travel to the Aldebaran star system.

He said that there were a staggering 1,442 individuals from all walks of life devoting resources to what could be likened to a modern day Noah's Ark to escape another prophesied World War:

Figure 23. Maria Orsic Photo which Meier Allegedly Saw and Claimed was Semjase

> Some of those people in the 1,442 were scientists and engineers in Germany, and some were shoe salesman. They all wanted to build a space ship. They borrowed money, took out loans on the farm, and were in debt up to their necks [as] they were trying to [buy] materials to build their spaceship. It was like they thought they could simply load their families and friends into their ship and fly off to some distant star.[45]

It may sound implausible that ordinary civilians could build vehicles, capable of deep space flight using teleportation principles, without significant assistance by national governments. However, in the early 1960's, Otis Carr, a protégé of Nikola Tesla, developed a flying saucer craft that used teleportation principles. Carr's goal was to also kickstart a civilian spacecraft industry. Unfortunately for Carr and his supporters, his civilian space program was terminated by government authorities who confiscated Carr's prototype spacecraft, the OTC-X1, and jailed him on concocted securities fraud.[46]

Figure 24. Otis Carr's OTC-X1 in 1961

The second flying saucer program in Nazi Germany was under the control of Himmler's SS, who were attempting to weaponize the program for the war effort. According to Tompkins, the spies reported that the Nazi's effort to win the war through weaponizing flying saucer technologies was being assisted by another group of extraterrestrials, called the Reptilians. This dangerous and conquest oriented faction had reached secret agreements with Hitler:

> The Navy agents (spies) in Germany discovered what all those "out of this world" aliens gave Hitler: UFOs, antigravity propulsion, beam weapons, extended life and plenty of mind-controlled willing girls programs. The Reptilians made a deal with the Third Reich SS giving them this big box full of toys in exchange for letting Hitler enslave the rest of the planet.[47]

Tompkins has further exposed how "Reptilian consultants" were helping the Nazi SS develop advanced weapons technologies:

> They had, if you want to call them, "consultants", who are Reptilian consultants assisting on all of these different things that it takes to design and build these spacecraft carriers, and propulsion systems. So this is an extremely well developed program and documented like crazy. Getting copies of the documents was hard for them, hard for our spies. This was an open program in the upper level of the SS.[48]

The spies learned that the goal of the Reptilian plan was not only to assist the Nazis to win the war and achieve planetary conquest, but to build fleets of antigravity spacecraft carriers that could be used for interplanetary conquest in other star systems:

> Holy cats the thing went way beyond that [world conquest]. Again, what we just said about this was the tip of the iceberg of what they were doing. Already Reptilians were doing it to other stars' planets all over this area of the Galaxy.... These young kids, the operatives, they couldn't believe half of what they brought back, what was going on. But some of them were really good people, and they knew how to get into places and listen to what's going on, and finding what that guy he had talked to, or he heard, and going to so and so, because yes they were doing that over there. [49]

Tompkins' information is startling, but it is not the first time claims have been made about Nazi Germany receiving extraterrestrial help during the War. The father of German rocket science, Herman Oberth, said the following in response to a question about Nazi Germany's rapid technological development:

> "We cannot take credit for our record advancement in certain scientific fields alone. We have been helped." When asked by whom, he replied, "the peoples of other worlds."[50]

In 1998, Linda Moulton Howe interviewed a former CIA agent (using the aliases Kewper and Stein), who from 1957 to 1960 was given access to highly classified briefing files. He said that the documents revealed that the Nazis had indeed been helped by an extraterrestrial race in their advanced aeronautical weapons projects:

> When Vril was building that first craft, the Vril had one or two of the aliens that worked with them in Germany where they fired rockets from Peenemunde. The documents I read in Washington said that's where the first Vril vehicle was made.... They [aliens] were helpers to Germany. [51]

More recently, another whistleblower, Corey Goode, says that from 1987 to 2007, he read intelligence briefings on smart-glass pads that detailed what had happened during World War II. Goode's claims concerning multiple "Secret Space Programs" he served on were extensively examined in my 2015 book, *Insiders Reveal Secret Space Programs and Extraterrestrial Alliances*, and found to be consistent with many historical documents, circumstantial evidence, and other whistleblower testimonies.

Goode described how German Secret Societies were helped, before and throughout World War II, by two different extraterrestrial groups. He identified one as a Reptilian-like race called the "Draconians" and the other as Nordics:

> The Germans were in contact and working hard alongside the Reptilians ... but there were Nordic groups involved as well that different elements of the German Space Program were in contact with. [52]

Both the Reptilians and Nordics were involved in assisting the Germans in the development of advanced antigravity spacecraft:

> At the same time these [German Secret] Societies had made contact with the Draco Federation and another group [Nordics] that avoided the Draco's. The German Occultists were very busy from the early 1900's, especially the time just before, during and after World War One. Their major breakthroughs occurred in the late 1930's.[53]

Goode said that the ultimate goal of the Reptilians, in helping German Secret Societies and the Nazi SS, was to create advanced space carrier battle groups that would be capable of interplanetary conquest. The eventual German/Nazi space battle groups became what Goode described as the "Dark Fleet." It operates outside of our solar system, and is discussed at length in the book, *Insiders Reveal Secret Space Programs and Extraterrestrial Alliances*.[54]

Consequently, the whistleblower testimonies by the CIA Agent and Corey Goode support Tompkins' claims that intelligence briefing files, dating from World War II, did indeed refer to Nazi Germany being assisted by extraterrestrials—two alien groups with very different agendas. The Nordics, or Aldebarans, had the objective to help Germany develop both technologically and spiritually for space exploration. They worked largely with private citizens/groups such as Orsic and the Vril Society, which Hitler allowed to continue once he came into power. In regard to this Tompkins commented in an interview:

> Germany found out about the blonde [Orsic], took her over, stopped everything, and then got to this point where there was some sort of pressurized program by the SS to control that original group. Now, several times they did work together, but Hitler allowed them to operate independently of the whole SS program – the whole development. So we had two developments going on in Germany. The girls didn't want their vehicles to be used for anything else but travel. They were afraid that somebody would get a hold of it and they'd use it for military, which is, of course, what they got.[55]

The Reptilians, on the other hand, had no ethical or spiritual program they wished to encourage among the Germans or on behalf of humanity, and worked primarily with the Nazi government to develop spacecraft which were essentially built to become weapons of war. The ultimate goal of the German-Reptilian alliance went far beyond winning World War II and mere planetary conquest, which Tompkins attests the US Navy had alarmingly learned from its spies.

Tompkins further says that in 1939, the Nazi SS was given the location of two large caverns in Antarctica by the Reptilians, and proceeded to move the bulk of the German secret space programs there over time. By August 1944, when it became clear that the war was lost after the successful Allied landing in Normandy, the Nazi's accelerated their efforts to relocate the best scientists, engineers and vital resources to Antarctica, and alternate Argentinian caverns, through specially built submarines capable of carrying very large cargoes.

According to Tompkins, the US Navy learned of the existence of these secret Antarctic bases directly from their spies embedded within Nazi Germany, who found that the Nazis/Germans used these remote bases to launch the first space missions by modern humanity. Elaborating on this, Tompkins says that the Germans began moving equipment and supplies to Antarctica as early as 1913. This coincides historically with the Second German Antarctic Expedition from 1911-1913 led by Wilhelm Filchner.[56] This period also corresponds with the increasing role of German secret societies in exploring remote global locations for occult knowledge. Significantly, Filchner later led Nazi expeditions to Nepal and Tibet. The movement of equipment to Antarctica accelerated in the lead up to World War II, as Tompkins highlights: "The move from Germany to Antarctica was in operation from ... [1938] way before the war started. In fact some of the stuff went down in 1913."[57]

Navy spies, who Tompkins heard candidly in debriefings, had learned about secret agreements which had been reached between Hitler's regime and Draconian Reptilians. Tompkins has revealed some of the major elements of this agreement and how it concerned Antarctica:

> Large portions of equipment were sent down there. But right next to them were three tremendous size caverns which the Reptilians had. Not Grays, but Reptilians. Germany got two more, about a tenth the size of the big Reptilians [cavern]. They were able to ... [go] down, usually

> by submarine. They built these flat submarines, these regular class, so they could ship all this stuff down.[58]

Tompkins' remarkable information is consistent with Grand Admiral Karl Donitz, who referred on three occasions to an impregnable fortress being built for Hitler in a remote location using Germany's advanced submarine fleet. In 1943, Donitz is reported to have stated: "the German submarine fleet is proud of having built for the Führer, in another part of the world, a Shangri-La on land, an impregnable fortress."[59]

The second occasion was in 1944, when he revealed how plans were in place to relocate Hitler so he could launch a new effort for his thousand-year Reich:

> The German Navy will have to accomplish a great task in the future. The German Navy knows all hiding places in the oceans and therefore it will be very easy to bring the Führer to a safe place should the necessity arise and in which he will have the opportunity to work out his final plans.[60]

Finally, Donitz's remarks at his Nuremberg war crime trial clearly suggest that it was Antarctica where Germany's most advanced technologies had been secretly relocated by his submarine fleet. At the trial he boasted of "an invulnerable fortress, a paradise-like oasis in the middle of eternal ice."[61]

Donitz's remarks were made plausible in 1966 by cartographer and artist for the National Geographic Society, Heinrich C. Berann. In Berann's depiction of an ice-free Antarctica, he shows underwater passageways that run throughout the Antarctic continent.[62] This provided a plausible way in which submarines could travel under the ice for considerable distances to Nazi Germany's "invulnerable fortress."

Figure 25. Illustration of Ice Free Antarctica by Heinrich Berann

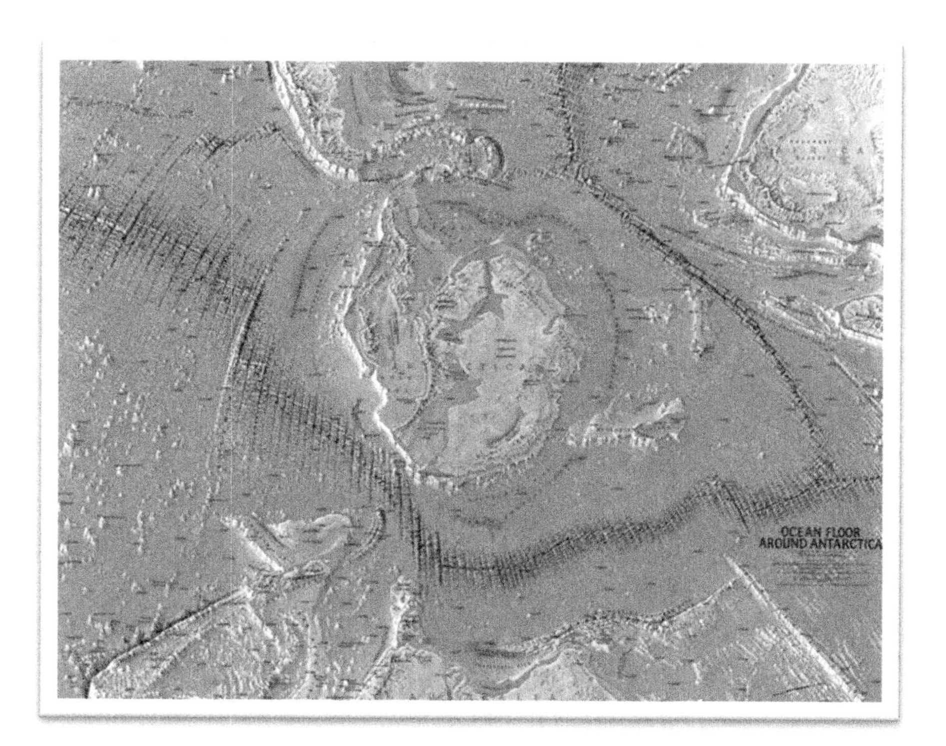

Donitz's claims are further supported by documents provided by an alleged German submarine crewman after the war, which described the instructions for U-Boat Captains to reach the Antarctica bases through the hidden passageways. Figure 26 displays an image of the document with the translated instructions. [63]

Tompkins describes simultaneous flying saucer programs that had been developed by the Nazis. One was in Nazi Occupied Europe, while the second was in Antarctica. In his response to a question about where Germany's antigravity craft were being built, he said:

Figure 26. Directions to Antarctic Bases

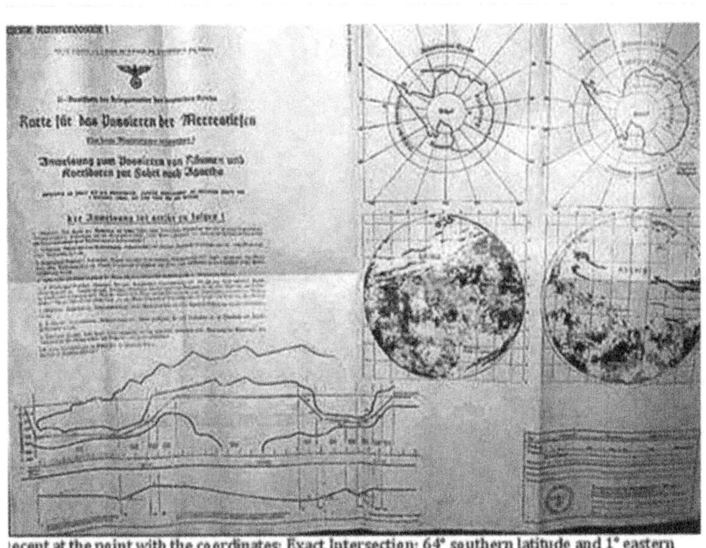

ecent at the point with the coordinates: Exact Intersection: 64° southern latitude and 1° eastern ongitude, to a depth of 400 meter.

he instruction have to be strictly followed!

1. Decent, from the point of decent with half speed, a starboard declination of 10° with a bow-heaviness declination angle of 5°. Distance 188 sm. Given depth – 500 meter. (Because of the moving inside the corridor the pressure on the ship body when maneuvering is insignificant)
2. Ascent. Full load with a stern trim. Ascent angle 23° with a port declination of 22°. 190 meter upwards. Distance 75.5 sm.
3. Difficult maneuver! Ascent full speed with a stern trim. Ascent angle 41°. Proceed straight ahaead. 110 meter upwards, distance 21.5 sm. Afterwards starboard declination of 8° until ascent to the surface in a distance of 81sm.
4. Proceed on the surface within the grotto with a starboard declination of 8°, Distance 286 sm.
5. 6. Schwieriges maneuver! Descent. With a bow-heaviness declination 45° to a depth of 240 meter. Distance 60 sm. Afterwards with a port declination of 20°, at which the descent to 310 meter to the entrance to the corridor continues. After the 310 meter mark the bow heavy descent need to be continued. Descent angle 7° until 360 meter, distance 70 sm. Futher starboard declination of 31° to a depth of 380 meter.
7. Descent. Bow heavy, ascent angle 22°, 100 meter upwards with a port declination of 26°. Distnace 43 sm.
8. Ascent. Stern trim. Ascent angle 45°, straight ahead until reaching the surface of Agartha. Distance 70 sm.
9. Proceed to Agartha. Full Speed. Proceed straight ahead, until the new light can be seen. Change of magnetic poles. The changes of the compass needle and instruments are to be disregarded.

(Further instructions in package Nr. 3 only when arrived in Agartha to be opened)

> They built the prototypes in Germany. They built pre-prototype, something which is ready for production, in Antarctica. They put this stuff in production in the countries all over Germany [Occupied Europe], and they continued to build similar vehicles in Antarctica.[64]

Next comes one of the most astonishing secrets gained by the Navy spies embedded in Germany. Tompkins shares that with the help of the Reptilians, the Nazi's Antarctica program had successfully launched manned missions to the Moon, planets, and even other star systems.

> Now the question was asked whether did we ever get to the Moon? You see it was in some way, it was well known that the Germans had a number of vehicles that flew out and came back. One of the first ones, they got into trouble, they crashed and the whole group died. But that was almost all coming from Antarctica. Way before, four years before, the war ended, they were always moving all of this stuff out. And so the flights, almost everything came from Antarctica with the same people. ... I don't know if it was true or not, but it was stated by some of those fellows [Navy spies] that they had gone to other stars and come back.[65]

Furthermore, Tompkins heard from the Naval operatives that the Germans had also achieved the first space flight to Mars in late April 1945, which had a crew of 30, including three Japanese astronauts. This stunning achievement ended in disaster when the spacecraft crash-landed, resulting in the death of the entire crew.

Tompkins remarkable claims again corroborate the testimony of secret space program whistleblower, Corey Goode, who says that the "smart glass pads" he had access to described the successful Nazi space program that operated out of Antarctica.

Goode stated, however, that the Antarctica program was controlled by German secret societies rather than the Nazi SS:

> As explained, the Nazi remnants that were made up mostly of Secret Societies that created a "Break Away Civilization," kept the most advanced technology secret from even their highest Military and Political leaders, setup enclaves in South America and Antarctica. The locations in Antarctica were some ancient civilization ruins that had remained occupied by certain groups in thermal areas that cause areas similar to lava tubes and domes under the glaciers.[66]

Goode goes on to corroborate Tompkins' claim that the Nazi bases were built adjacent to caverns controlled by another advanced civilization, the Reptilians:

> There was an underground and under glacier city complex that was already occupied and setup in a couple of locations and the NAZI's renovated an area that was mostly crushed above the surface but had plenty of room under the domed ice, thermal underground energy and caverns (accessible via U-boat under the ice flows and openings that made it ideal for a hidden multipurpose base) that were perfect for them to secretly build out during the entire Second World War.[67]

Statements by Admiral Donitz support Tompkins' claims that Nazi Germany had succeeded in locating and building underground facilities under the Antarctic continent. The long route under the Antarctic ice sheets, which the German submarine fleet allegedly took to reach these hidden caverns, gives credence to Tompkins' testimony that Reptilian extraterrestrials had provided the Nazis with the information necessary to locate the hidden Antarctic caverns, and the under-ice passageways to reach them.

What the Navy spies were revealing to Admiral Botta, and his subordinates, which included Tompkins at the Naval Air Station from 1942 to 1946, was astonishing. The information made it even more urgent that what was retrieved by the Navy and Army from the Los Angeles Air Raid be investigated for its potential war applications, as well as bringing all captured Nazi prototypes from liberated Europe to the US for further study:

> At the end of World War II, Naval Intelligence operators (spies) penetrated virtually every German secret weapons, advance system, rockets, aircraft, UFO's and heavy water [projects] in the country. They located the individuals in these facilities, and they were tagged. When the hostilities ceased, the Naval Intelligence and additional intelligence officers went straight into these locations and removed not only the research scientists, but their documentation, and as much of the weapons system as they could. They were all brought to the United States in what was called Project Paperclip.[68]

The twin imperatives of learning about extraterrestrial and Nazi flying saucer technologies led to an informal working group, initiated by Donald Douglas, founder of the Douglas Aircraft Company, evolving into Project RAND by the end of World War II.

Endnotes

[25] These documents are available online at The Mussolini UFO files: http://web.tiscali.it/lareteufo/mussof1.htm (accessed 6/30/15).
[26] For detailed discussion of the Italian UFO study group created in 1933, see Michael Salla, *Insiders Reveal Secret Space Programs and Extraterrestrial Life* (Exopolitics Institute, 2015) pp. 83-104.
[27] Alfredo Lissoni, "New Documents 'Will Revolutionize UFOlogy'! (UFO Cover-Up By Mussolini)" http://www.ufoevidence.org/documents/doc1885.htm (accessed 7/29/15).
[28] Andrew Gumbel, "Scholars in a spin over Churchill link to the death of Mussolini," *The Independent*, http://www.independent.co.uk/news/world/scholars-in-a-spin-over-churchill-link-to-the-death-of-mussolini-1601820.html
[29] "A Lesson in Naval History in 1:600 Scale," http://craftsmanshipmuseum.com/Tompkins.htm
[30] "A Lesson in Naval History in 1:600 Scale," http://craftsmanshipmuseum.com/Tompkins.htm
[31] ExoNews TV, "Interview Transcript – Navy Disseminated Nazi Antigravity Secrets to Leading US Companies & Think Tanks," http://exopolitics.org/interview-transcript-navy-disseminated-nazi-antigravity-secrets-to-leading-u-s-companies-think-tanks/
[32] Transcript from video recording February 25, 2016. See Apendix 1.
[33] Transcript from video recording February 25, 2016. See Apendix 1.
[34] William Tompkins, *Selected by Extraterrestrials*, p. 427.
[35] http://www.ww2f.com/topic/29345-amm2-duties/?p=361199
[36] I am very grateful to Duke Brickhouse, J.D., who drafted and sent the Freedom of Information Act letters on my behalf to the US National Archives to gain the release of Admiral Rico Botta's personnel files.
[37] Email from Dr. Robert Wood, September 25, 2016.
[38] FBI, The Vault, https://vault.fbi.gov/UFO/UFO%20Part%202%20of%2016/view#document/p5
[39] Tom Carey and Donald Schmitt, *Inside the Real Area 51: The Secret History of Wright Patterson* (New Page Books, 2013) Kindle Edition Location 329.
[40] See http://exopolitics.org/wp-content/uploads/2016/07/Admiral-Botta-Retroactive-Promotion.jpg Also available at: http://www.senate.gov/legislative/LIS/executive_calendar/1946/05_25_1946.pdf

[41] Ken McCanliss, "Retroactive Promotions to Flag Officer," http://forum.axishistory.com/viewtopic.php?t=14146&start=15#p721732
[42] Private interview January 16, 2016.
[43] Justin Deschamps, "David Wilcock and Corey Goode: History of the Solar System and Secret Space Program - Notes from Consciousness Life Expo 2016 ," http://sitsshow.blogspot.com.au/2016/02/david-wilcock-and-corey-goode-history.html
[44] For more about Marconi's flying saucer program in South America, see Michael Salla, *Insiders Reveal Secret Space Programs and Extraterrestrial Alliances,* pp. 225-27.
[45] William Tompkins, *Selected by Extraterrestrials*, p. 427.
[46] A chapter about Otis Carr's civilian space craft program appears in *Insiders Reveal Secret Space Programs and Extraterrestrial Life*, pp. 225-48.
[47] William Tompkins, *Selected by Extraterrestrials*, pp. 70-71,
[48] William Tompkins, "Interview Transcript – US Navy Spies Learned of Nazi Alliance with Reptilian Extraterrestrials," *ExoNewsTV,* http://exopolitics.org/interview-transcript-us-navy-spies-learned-of-nazi-alliance-with-reptilian-extraterrestrials/
[49] William Tompkins, "Interview Transcript – US Navy Spies Learned of Nazi Alliance with Reptilian Extraterrestrials," *ExoNewsTV,* http://exopolitics.org/interview-transcript-us-navy-spies-learned-of-nazi-alliance-with-reptilian-extraterrestrials/
[50] Robin Collyns, *'Did Spacemen Colonize the Earth?'* (Pelham Books, 1974) 236. "The Aldebaran Mystery: The Nazi/ET UFO Connection," http://ufodigest.com/news/0208/aldebaran-mystery.html (accessed 10/29/16).
[51] Interviewed by Linda Moulton Howe, Earth Files, http://www.earthfiles.com/news.php?ID=1464&category=Real%20X-Files (accessed 6/6/15).
[52] Interview Transcript, "Cosmic Disclosure (S5E10): SSP Testimonials with William Tompkins," http://spherebeingalliance.com/blog/transcript-cosmic-disclosure-ssp-testimonials-with-william-tompkins.html
[53] Cited by Michael Salla, "Corporate bases on Mars and Nazi infiltration of US Secret Space Program," http://exopolitics.org/corporate-bases-on-mars-and-nazi-infiltration-of-us-secret-space-program/
[54] Michael Salla, *Insiders Reveal Secret Space Programs and Extraterrestrial Alliances, pp. 117-46.*
[55] Interview Transcript, "Cosmic Disclosure (S5E10): SSP Testimonials with William Tompkins," http://spherebeingalliance.com/blog/transcript-cosmic-disclosure-ssp-testimonials-with-william-tompkins.html

---

[56] See Wikipedia, https://en.wikipedia.org/wiki/Wilhelm_Filchner
[57] William Tompkins, "Interview Transcript – Reptilian Aliens Helped Germans Establish Space Program in Antarctica," http://exopolitics.org/interview-transcript-reptilian-aliens-helped-germans-establish-space-program-in-antarctica/
[58] William Tompkins, "Interview Transcript – Reptilian Aliens Helped Germans Establish Space Program in Antarctica," http://exopolitics.org/interview-transcript-reptilian-aliens-helped-germans-establish-space-program-in-antarctica/
[59] "Antarctic Enigma," http://www.bibliotecapleyades.net/tierra_hueca/esp_tierra_hueca_6c.htm (accessed 10/29/16).
[60] "The Antarctic Survival Myth," http://www.bibliotecapleyades.net/antarctica/antartica22.htm
[61] "The Antarctic Survival Myth," http://www.bibliotecapleyades.net/antarctica/antartica22.htm
[62] "Ancient Antarctica Was As Warm As Today's California," http://atlanteangardens.blogspot.com/2014/04/ancient-antarctica-was-as-warm-as.html
[63] "Ancient Antarctica Was As Warm As Today's California," http://atlanteangardens.blogspot.com/2014/04/ancient-antarctica-was-as-warm-as.html
[64] William Tompkins, "Interview Transcript – Reptilian Aliens Helped Germans Establish Space Program in Antarctica," http://exopolitics.org/interview-transcript-reptilian-aliens-helped-germans-establish-space-program-in-antarctica/
[65] William Tompkins, "Interview Transcript – Reptilian Aliens Helped Germans Establish Space Program in Antarctica," http://exopolitics.org/interview-transcript-reptilian-aliens-helped-germans-establish-space-program-in-antarctica/
[66] Cited by Michael Salla, "Corporate bases on Mars and Nazi infiltration of US Secret Space Program," http://exopolitics.org/corporate-bases-on-mars-and-nazi-infiltration-of-us-secret-space-program/
[67] Cited by Michael Salla, "Corporate bases on Mars and Nazi infiltration of US Secret Space Program," http://exopolitics.org/corporate-bases-on-mars-and-nazi-infiltration-of-us-secret-space-program/
[68] William Tompkins, *Selected by Extraterrestrials*, p. 139.

CHAPTER 3

PROJECT RAND & THE FUTURE OF NON-TERRESTRIAL TECHNOLOGIES

Leaked "Majestic Documents" reveal that President Roosevelt, following the advice of military officials, decided that the advanced science behind the recovered antigravity craft from the Los Angeles Air Raid would be too challenging to reverse engineer in a timeframe suited to the war effort. In a 1944 memorandum to his science advisor, Dr. Vannevar Bush, Roosevelt firmly ruled out a research and development program of "non-terrestrial" technologies based on antigravity principles. Expense was the issue and it would impact on the development of more conventional military weapons such as the atomic bomb. In this Top Secret memorandum, Roosevelt wrote:

> Various points have been raised about the difficulties such an endeavor would pose to the already hardened research for advanced weapons programs and support groups in our war effort and I agree that now is not the time. It is my personal judgment that, when the war is won, and peace is once again restored, there will come a time when surplus funds may be available to pursue a program devoted to understanding non-terrestrial science and its technology which is still greatly undiscovered.[69]

Yet, Roosevelt went on to postulate that the US would eventually play a leading role in developing the incredible technologies that had fallen into their hands.

Figure 27. Top Secret Document Signed by President Roosevelt

DOUBLE TOP SECRET — February 22, 1944.

THE WHITE HOUSE

WASHINGTON

February 21, 1944

MEMORANDUM FOR

THE SPECIAL COMMITTEE ON NON-TERRESTRIAL SCIENCE AND TECHNOLOGY

[illegible] agree with the O [illegible] D proposal of the recommendation put forward by Dr. Bush and Professor Einstein that a separate program be initiated at the earliest possible time. I also agree that application of non-terrestrial know how in atomic energy must be used in perfecting super weapons of war to affect the complete defeat of Germany and Japan. In view of the cost already incurred in the atomic bomb program, it would, at this time, be difficult to approve without further support of the Treasury Department and the military. I therefore have decided to forego such a enterprise. From the point of view of the informed members of the United States, our principle object is not to engage in exploratory research of this kind but to win the war as soon as possible.

Various points have been raised about the difficulties such an endeavor would pose to the already burdened research for advanced weapons programs and support groups in our war effort and I agree that now is not the time. It is my personal judgement that, when the war is won, and peace is once again restored, there will come a time when surplus funds may be available to pursue a program devoted to understanding non-terrestrial science and its technology which is still greatly undiscovered. I have had private discussions with Dr. Bush on this subject and the advice of several eminent scientists who believe the United States should take every advantage of such wonders that have come to us. I have heard the arguments of General Marshall and other members of the military that the United States must assume its destiny in this matter for the sake of the Nation's security in the postwar world and I have given assurances that such will be the case.

I appreciate the effort and time spent in producing valuable insights into the proposal to find ways of advancing our technology and national progress and in coming to grips with the reality that our planet is not the only one harboring intelligent life in the universe. I also commend the committee for the organization and planning that is evident in Dr. Bush's proposal and the delicate way in which it was presented. I trust the committee will appreciate the situation on which this office must render its decision.

[illegible]

DOUBLE TOP SECRET

Consequently, only after the war would significant scientific resources finally be devoted to research and development of the recovered antigravity craft. In the meantime, the US military would gather as much intelligence as possible on what the Axis Powers were doing in this arena, and endeavor to further retrieve any craft of non-terrestrial origin that came their way.

During this interim, at least three things were done to gain necessary intelligence data in preparation for the time when the US government and its military could channel significant resources into developing a comprehensive scientific plan for the flying saucer phenomenon. First, the Army Air Force, through the Interplanetary Phenomenon Unit, would gather as much intelligence as it could on antigravity vehicles, and relocate any craft that came into US military possession, or that of its allies, to the United States.

Second, the Navy, through its covert operatives in Nazi Occupied Europe, would gather data secretly on what the Nazis were doing in their research and development of flying saucer technologies. And third, preliminary studies would be conducted by an informal study group capable of fully appreciating the military and scientific potential of what had been acquired from the Los Angeles Air Raid, along with the intelligence provided by both the Army Air Force and the Navy. The informal study group would use the right people and protocols to prepare for the future when funding and personnel became available for a comprehensive reverse engineering program of the captured flying saucers and the data retrieved from Nazi Germany. William Tompkins has disclosed key information about this informal study group.

Shortly after the 1942 Los Angeles Air Raid incident, Tompkins says the President of Douglas Aircraft Company, Donald Douglas Sr., along with his chief engineer, Arthur Raymond, and his assistant, Franklin Colbohm, convened an informal working group that included two Army Air Force Generals and two Navy admirals. The goal was to investigate the scientific principles and research the implications of the two retrieved UFO craft.[70]

The Douglas Aircraft Corporation was headquartered in Santa Monica, so it is natural to assume that senior Douglas personnel would have witnessed the Los Angeles event. An immediate investigation began, which is certainly what Tompkins contends in his autobiography, where he wrote:

> Frank Collbohm – under the direction of Arthur Raymond and Donald Douglas – has been secretly investigating the strange flight of vehicles over Santa Monica and Los Angeles since 1942.[71]

In addition, Tompkins described the powerful influence the Los Angeles incident had on Donald Douglas and other key individuals who directly participated in or became aware of the informal working group, which came to be known in-house as Advanced Design. This group would ultimately lead to the formal establishment of Project RAND in October 1945:

> There were several people in the Navy Department, Army Air Corps, and an aircraft company whose entire lives were impacted. They were Admiral Roscoe H. Hillenkoetter, Secretary of the Navy, James V. Forrestal, Army Air Corps General Nathan F. Twining and General Curtis Le May, Edward Bowles of M.I.T., Dr Vannevar Bush and Donald Douglas, Sr., to name a few.[72]

Douglas Aircraft Company was among the leading manufacturers of aircraft for both the US Navy and the US Army Air Force at the beginning of World War II. By 1945, Douglas produced nearly 30,000 aircraft for the successful war effort, and its engineers and production facilities were world-renowned. Tompkins was very familiar with the operations and history of Douglas Aircraft Company, due both to his employment in its engineering division from 1950 to 1963, and also because of his visits to Douglas when working under Rear Admiral Rico Botta out of the San Diego Naval Air Station from 1942 to 1946.

Consequently, it is very plausible that Donald Douglas Sr. would have contacted senior officers in both the Army and Navy in early 1942, to form an informal working group to discuss the Los Angeles UFO incident, and its implications for the aeronautical industry. After all, Douglas Aircraft Company was ramping up aircraft production for both the Army Air Force and Navy for the war effort, and it behooved everyone to learn as much as possible about the UFOs involved in the Los Angeles incident.

In the immediate months after the Air Raid took place, intelligence data about the Nazi research and development programs began to accumulate at Naval Air Station, San Diego, and was made known to Douglas Aircraft Company according to Tompkins. In the previous chapter, a document was presented showing Admiral Botta traveling to Douglas Aircraft Company in October 1942 to discuss experimental aircraft issues. This was likely the first time Douglas would have been told about the Navy's gathering of intelligence on the Nazi programs.

After beginning his covert assignment under Botta, Tompkins says that from the Spring of 1943 onward, he travelled at least three times to Douglas Aircraft Company with the briefing packets containing the latest information about the Nazi programs.

> I had been flying up to Douglas, from Naval Air Station North Island, San Diego, since spring 1943. I was staff to Admiral [Rico Botta], Commander, Naval Intelligence. I had his high wing DH-2 aircraft assigned to me as needed for these and other local, western [U.S.A.] missions.... visited classified Douglas Santa Monica and Cal Tech facilities, 3 to 4 times each. [73]

Tompkins named both Arthur Raymond and Franklin Colbohm as being directly involved in the study of the briefing packets delivered to Douglas Aircraft Company. In an interview,

Tompkins recalls that one of the Douglas engineers he spoke to about the briefing packets may have been Colbohm, who later became the first President of the Rand Corporation.[74] Consequently, Tompkins' claim that the genesis of this informal working group, established by Donald Douglas, would lead three years later to the formal establishment of Project RAND in October 1945, is supported by historical events.

Tompkins described the formation and purpose of Project RAND and its connection to a top secret think tank in the engineering division of Douglas as follows:

> The Douglas/RAND scientists and concept conceivers were studying topics actually way above top secret. They were in a Think Tank deep inside the Douglas Aircraft Company's Engineering Department A-250, that nobody knew existed at the Santa Monica, California Airport.[75]

He identified some of the key people involved in the Project RAND/Douglas initiative:

> Under the direction of James Forrestal, who was the Secretary of the Navy, on October 1, 1945, several high-ranking big shots were brought on board, including General Hap Arnold, Edward Bowles (of M.I.T.) and a consultant to the Secretary of the War, Donald Douglas ... They met in secret at the Army Air Corps [Force] Headquarters, Hamilton Field, California to set up Project RAND, a way-above-top-secret scientific think tank. It was created in ... [October] 1945, as a special contract to Douglas Aircraft Company. At the Santa Monica Municipal Airport. Inside a highly classified walled-off area in the Douglas Engineering Department, **Project RAND studied the implications of threating alien agendas**.... Then, on March 2, 1945, a letter of contract was executed, which put Project RAND under the direction of Douglas's Assistant Chief Engineer, Frank

> Collbohm. The Douglas Think Tank was born. [emphasis added][76]

Tompkins emphasized the exclusive role given to the Douglas campaign at the end of the war in establishing the first US think tank set up to focus on flying saucer craft, and the possible threat posed by extraterrestrial visitors as starkly exemplified by the 1942 Los Angeles Air Raid incident:

> Unknown to most of us in 1945, the Douglas Aircraft Company was "sole-source selected." It was given an above top secret RAND contract to study and locate military satellites in earth orbit (unofficially, alien threats). The Advanced Design Section in Engineering was extensively expanded to accommodate the massive problem. The first galactic Think Tank on this planet was formed. [77]

Further, Tompkins explained how the Project RAND contract gave Douglas employees access to the technical information possessed by all other US research and development programs that were relevant to flying saucer technology and extraterrestrial life:

> This contact provided almost complete access to, and support from, nearly all of the technical individuals and organizations in the U.S.... This RAND contact thrust Douglas into fantastic, technical programs. They defined the alien threats, and researched every possible method and technical task for the people involved in Naval defense and offensive missions. They were methods designed to conceive Naval space missions and battle groups, and to design space vehicles/spacecraft carriers and weapons to combat the alien battle groups. It was to be supported technically North American Aviation, Inc, Northrup Aircraft Company, Lockheed Aircraft Corporation, Jet Propulsion

> Laboratory, SRI, MIT and Caltech. What we knew was dwarfed by all we had to learn. [78]

A document that directly supports Tompkins' claim that Project RAND was primarily set up to study the flying saucer phenomenon, and collaborated with other leading US scientific organizations conducting advanced aerospace research in doing so, is one of the leaked "Majestic documents." The "White Hot Report" lists Project RAND among the research organizations studying artifacts recovered from the crash of extraterrestrial vehicles:

> Based on all available evidence collected from recovered exhibits currently under study by AMC, AFSWP, NEPA, ABC, NACA, JRDB, **RAND,** USAAF, SAG and MIT, are deemed extraterrestrial in nature. [emphasis added][79]

Figure 28. Leaked "White Hot Report" dated Sept 19, 1947. Source: Majestic Documents

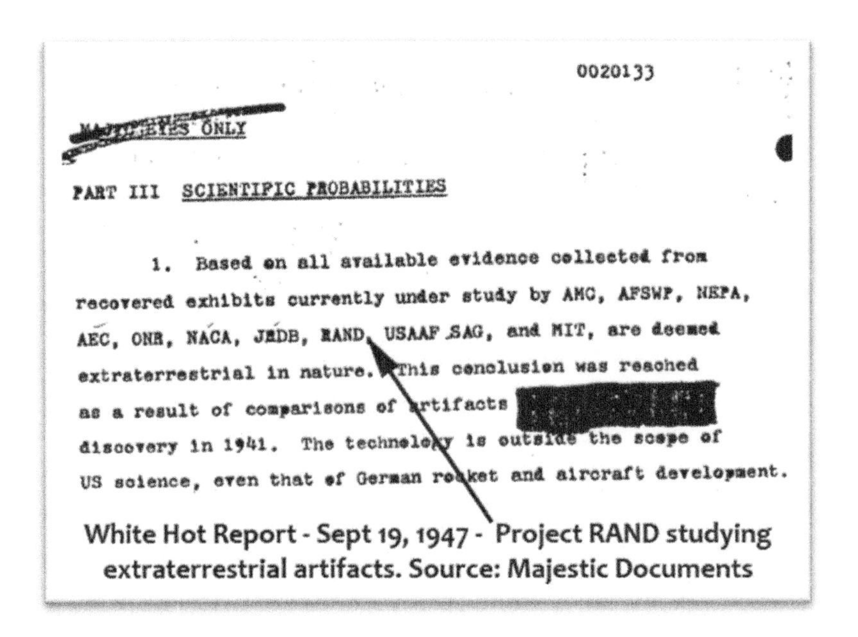

0020133

~~MAJIC EYES ONLY~~

PART III <u>SCIENTIFIC PROBABILITIES</u>

1. Based on all available evidence collected from recovered exhibits currently under study by AMC, AFSWP, NEPA, AEC, ONR, NACA, JRDB, RAND, USAAF SAG, and MIT, are deemed extraterrestrial in nature. This conclusion was reached as a result of comparisons of artifacts [redacted] discovery in 1941. The technology is outside the scope of US science, even that of German rocket and aircraft development.

White Hot Report - Sept 19, 1947 - Project RAND studying extraterrestrial artifacts. Source: Majestic Documents

Significantly, the leaked "White Hot Report" received the highest level of authenticity from the independent investigation by veteran document researchers, Dr. Robert Wood and Ryan Wood.[80] Consequently, the "White Hot Report" document is independent confirmation that the Douglas Aviation Company, through Project RAND, was involved in the study of retrieved alien spacecraft, just as Tompkins claimed. This also supports another core element of Tompkins' testimony regarding him joining a secret engineering division think tank in Douglas in 1951, which did feasibility studies on different antigravity spacecraft for the Navy.

Tompkins told the following about the secrecy surrounding Project RAND: "They had the highest, secret clearance even above the nuclear bomb." [81] His statement about the extraordinary level of security given to Project RAND is supported by a document written by a senior radio engineer with the Canadian Department of Transportation. Dated November 21, 1951, Wilbert Smith wrote to the Controller of Telecommunications concerning flying saucer technologies being secretly studied in the US at the time: "The matter is the most highly classified subject in the United States government, rating higher than the H-bomb."[82] A November 1965 article in *The Progressive*, by journalist Wesley Marx, further corroborates Tompkins' claim about the extraordinary security procedures set in place by the Douglas Company for Project RAND:

> Douglas furnished administrative services, security guards, and locked rooms in its Santa Monica, California, facility, and RAND became a subsidiary, but virtually autonomous, division of Douglas.[83]

In addition to Donald Douglas, another one of the founders of Project RAND was the Commanding General of the Army Air Force, Henry H. "Hap" Arnold. He wrote a report to General Marshall, the Army Chief of Staff, stating:

> During this war the Army, Army Air Forces, and the Navy have made unprecedented use of scientific and industrial resources. The conclusion is inescapable that we have not yet established the balance necessary to insure the continuance of teamwork among the military, other government agencies, industry, and the universities. Scientific planning must be years in advance of the actual research and development work.[84]

Project RAND was set up to provide a think tank to help the military services develop the necessary strategic scientific planning that would be essential for the eventual research and development of advanced aerospace technologies, especially those using antigravity and other exotic propulsion systems. This was vitally influential to developing long term plans when it came to UFOs that were either of interplanetary origin, or linked to Nazi advanced technology projects.

Significantly, Arnold was also a key figure in the creation of the Interplanetary Phenomenon Unit set up to deal with UFO crashes soon after the Los Angeles incident. At the time, Arnold was the Commanding General of the newly created Army Air Force, which had been formed only eight months earlier in June 1941, and unified the two previously separate entities, Army Air Corps and General Headquarters Air Force. As the Commanding General of the Army Air Force, Arnold would have had to provide personnel with the necessary technical expertise for the secretive Army intelligence unit specially created to investigate reports of aircraft crashes that involved interplanetary UFOs or other forms of advanced aerial technologies.

Arnold's involvement establishes an important connection between the formation of Project RAND and the Interplanetary Phenomenon Unit. This supports Tompkins' claim that Project RAND emerged out of an informal study group to examine the Los Angeles UFO incident in 1942, and that from the beginning RAND's secret mission has been to help develop a long term coordinated policy on the UFO phenomenon and extraterrestrial life.

In his November 1965 article, Marx described how Arnold raised money for Project RAND, bypassing Congress and the conventional bidding process altogether:

> Without Congressional authorization and without taking bids, Air Force General H.H. "Hap" Arnold managed to shift around enough funds to award a $10 million contract to Douglas Aircraft Company to set up Project RAND.[85]

Ten million dollars, in 1945 terms, converts to as much as 790 million dollars in 2016 terms.[86] At the time, this was an enormous amount of money to give to an aviation company to conduct research, without taking bids from competitors. Arnold's decision shows that at that key time, he believed that the Douglas Company was well ahead of its main rivals, Lockheed, Boeing and Northrup, when it came to research and development of advanced aviation technologies.

Arnold's choice of the Douglas Company for funding Project RAND in 1945 supports Tompkins' claim that an informal study group had been formed at Douglas in 1942, after the Los Angeles Air Raid incident. By 1945, Douglas had become the obvious private entity to develop a long term coordinated plan between the government, military and private sector when it came to the study of flying saucer and antigravity technologies.

In addition to General Arnold, the RAND website identifies others involved in setting up the RAND Project:

> Other key players involved in the formation of this new, private organization were Major General Curtis LeMay; General Lauris Norstad, Assistant Chief of Air Staff, Plans; Edward Bowles of the Massachusetts Institute of Technology, consultant to the Secretary of War; Donald Douglas, president of the Douglas Aircraft Company; Arthur

> Raymond, chief engineer at Douglas; and Franklin Collbohm, Raymond's assistant. The name of the organization? Project RAND.[87]

The historical overview found in the website of the RAND Corporation, the successor to Project RAND, is notable since it confirms the leading role of the Douglas Aircraft Company in supporting the US military in helping to develop the necessary scientific plans for researching and developing advanced aerospace technologies. The expertise of Douglas' scientists and engineers was unparalleled at the time.

While the RAND Corporation website corroborates Tompkins' claims of leading Army Air Force generals and Douglas Aircraft Company engineers being involved in the formation of Project RAND, there is no mention of the Navy or of Secretary Forrestal's involvement. Why?

Funding for Project RAND was provided by the Army Air Force, through General Arnold, as noted earlier. The RAND contract came to be managed by the newly created Army position of "Deputy Chief of Air Staff for Research and Development," who in October 1945 was General Curtis LeMay—also an acknowledged co-founder of Project RAND. LeMay was in charge of the Army Air Force research facilities at Wright Field, Dayton, Ohio, which had received the retrieved flying saucer craft from the Los Angeles Air Raid incident. Wright Field would also receive other craft retrieved by the Interplanetary Phenomenon Unit, including the July 1947 Roswell crash, and also Nazi antigravity craft brought to the United States under Operation Paperclip.

The fact that only Army Air Force officials would be acknowledged, by official RAND historians, in its establishment was a reflection of the principal funder being the Army. It is also due to the Navy's downplaying of its wartime interest in flying saucer and antigravity research, as exemplified by Secretary of the Navy Frank Knox's public dismissal of the Los Angeles Air Raid incident as a false alarm. Yet the Navy, according to Tompkins, immediately began working behind the scenes with the Army Air Force and Douglas Aircraft engineers, through the Douglas informal study group from 1942 to 1945, to understand the phenomenon. At the same time, the Navy ran its covert intelligence program with operatives embedded in Nazi Germany's flying saucer programs.

Quickly, the Navy realized that the Army Air Force planned to assert its primacy when it came to the newly emerging field of space operations. Put simply, the Army Air Force viewed space as turf belonging to the soon to be created US Air Force (1947). Through Project RAND, scientific and military study of the flying saucer phenomenon would serve to exclude the Navy from playing a leading role in space. A policy battle arose between the Army Air Force and Navy when it came to developing the first system of space surveillance satellites. This policy battle is exemplified in a brief description on the RAND Corporation website:

> Interest in the feasibility of space satellites had surfaced somewhat earlier in a Navy proposal for an interservice space program (March 1946). Major General Curtis E. LeMay, then Deputy Chief of the Air Staff for Research and Development, considered space operations to be an extension of air operations.[88]

LeMay commissioned a study by Project RAND that was published in May 1946, titled "Preliminary Design of an Experimental World-Circling Spaceship."[89] The study led to the Navy losing its emerging policy struggle with the Army Air Force over jointly developing a space satellite reconnaissance system. Subsequent RAND publications then served to create the impression that only the Army Air Force, and later the US Air Force (formed September 1947), was serious about space operations, and in developing the exotic propulsion systems needed for space flight.

In the meantime, the Navy would secretly continue its own research and development work on reverse engineering the recovered extraterrestrial craft from the Los Angeles Air Raid, and the Nazi flying saucer prototypes acquired at the end of the war that were brought to its own research facilities. The principal location the Navy would choose to conduct its examination of flying saucers technologies was to be its flagship research and development location in California, the Naval Air Weapons Station, China Lake.

Tompkins says that during World War II, he flew multiple missions to China Lake with the briefing packets he had prepared from the debriefings of Navy spies: "Spent much time at China Lake facility modifying rockets for testing (40 visits) and working on their alien projects."[90] This reflects the Navy's sustained interest in understanding extraterrestrial technologies after the recovery of the alien craft from the Los Angeles incident, and its broader goals for future Navy space operations.

Due to Project RAND, all of the general public's attention fell upon the Army Air Force when it came to future space operations and the UFO phenomenon. This escalated after the July 1947 Roswell Incident when the Army Air Force announced that a flying saucer had crashed, only to retract the announcement hours later.[91] The effect of giving the public the impression that the Army Air Force was covering up the flying saucer phenomenon was laid. In this way, public attention was firmly directed away from what the Navy was secretly doing at its leading research and development facilities.

By late 1947, a decision was made to formally separate Project RAND from Douglas, which the RAND Corporation website explains:

> By late 1947, it seemed as though Project RAND—which was already operating fairly autonomously—should consider separating from Douglas. In February 1948, the Chief of Staff [Carl A. Spaatz] of the newly created United States Air Force wrote a letter to the president of the Douglas Aircraft Company that approved the evolution of Project RAND into a nonprofit corporation, independent of Douglas.[92]

According to Tompkins, the separation was not as smooth a transition as the RAND website depicts. In an interview, he said that roughly two thirds of the approximate 200 personnel working in Project RAND moved over to the Santa Monica facility of the newly created RAND Corporation. The one third left remained with Douglas at its own highly classified engineering facility in Santa Monica.[93]

It is this remnant of Project RAND that stayed with the secret Douglas think tank, Advanced Design, which Tompkins began working for in early 1951, after joining Douglas only months earlier in late 1950. Tompkins likened the separation of Project RAND from Douglas as a traumatic divorce, as he describes in his autobiography:

> I had been flying up to Douglas, from Naval Air Station North Island San Diego, since spring 1943... Now seven years later, it had been over a year since the divorce. The atmosphere in the Tank was thick with overwhelming tension and resentment. The personnel who had been forced to stay with the family [Douglas Aircraft Company] were hit the hardest. Those who wanted to stay, but who had to leave, also had strong feelings.[94]

Tompkins went on to explain his surprise when he learned in 1951 of the antigravity projects being secretly studied by the Douglas scientists and engineers in its classified think tank, and how a separate think tank [RAND Project] also existed with a similar purpose:

> We were totally unaware that this strange, out-of-this-world thing had ever existed inside a classified engineering area, inside an aircraft company, one which in turn, must have existed inside another classified area. Another Think Tank [RAND], comprised of other people, was also tasked to evaluate the unbelievable events.[95]

Although Advanced Design continued, albeit restructured, with the nearly 70 scientists and engineers remaining from Project RAND, an antagonistic relationship quickly developed between the RAND Corporation and the Douglas think tank. What Tompkins says next in his book illustrates that this became a long term problem:

> For years, this was a thorn in our sides. We were subjected to continuous interruptions in our conference rooms, by other people who appeared there to evaluate our studies and concepts.[96]

These "other people" were agents of industrial sabotage, according to Tompkins, and they would set back the Navy's efforts to work with Douglas' Advanced Design think tank towards the goal of creating antigravity vehicles.

Extraterrestrial Assistance & Industrial Sabotage

Advanced Design would find itself to be a player in two unfolding alien agendas. Put briefly, on the one hand, human-looking extraterrestrials that looked Scandinavian, and hence described by Tompkins as "Nordics," began helping Douglas think tank engineers to design kilometers-long space carriers and cruisers.[97] Tompkins says that the Nordic extraterrestrials were cultivating the Navy as a future ally because of the military conflicts that were occurring with (Draconian) Reptilians, elsewhere in the galaxy:

> I'm convinced some of those white hat aliens [Nordics] are pushing us. They want us to develop spacecraft carriers and help their Naval space battle groups combat the black hat aliens [Reptilians].[98]

On the other hand, Tompkins explains further, the Reptilians, along with their Nazi allies, were sabotaging projects that Douglas Advanced Design was working on for a future Navy-run space program. The RAND Corporation, according to Tompkins' testimony, was linked to this industrial sabotage, suggesting that RAND and the USAF were being influenced by the Reptilian extraterrestrials and Nazis.

There are two major issues to consider here when evaluating Tompkins' incredible statements about Nordic and Reptilian extraterrestrials using the US Navy and Air Force as proxies in a galactic conflict. Both directly relate to the history of the RAND Corporation and its current activities.

First, it is well known that former Navy Secretary, James Forrestal, who became the inaugural Secretary of Defense in September 1947, was locked in a number of bitter policy struggles with Stuart Symington, the first Secretary of the US Air Force (from September 1947 to April 1950).[99] What is not well known, according to Tompkins, is that Forrestal was in charge of the Navy espionage program on the Nazi antigravity programs and their alliance with Draconian-Reptilians during World War II. Tompkins says that the findings of the Navy espionage program led directly to Forrestal's order to establish the Office of Research and Inventions in May 1945 (see Figure 29), which formalized his oversight of the covert intelligence program run by Admiral Botta out of Naval Air Station, San Diego.

Forrestal's role in the Navy's espionage program was very likely a significant factor in his promotion to Secretary of Defense in September 1947. Certainly he was fully aware of the nuances and extent of the looming problem. Forrestal was in favor of the Navy taking a leadership role in solving the Nazi and extraterrestrial problem, and to reveal more of the truth to the general public.

Tompkins states that Forrestal was being directly influenced by Nordic extraterrestrials who had made contact with him. In an

Figure 29. Forrestal Creates Office of Research and Invention

(?/PHI:hg/vc 19 May 1945

To: All Bureaus, Boards and Offices of the Navy Department
The Commandant, U. S. Marine Corps
The Commandant, U. S. Coast Guard

Subj: Office of Research and Inventions.

1. There is hereby established, in the Office of the Secretary of the Navy, the Office of Research and Inventions at the head of which shall be a Chief and an Assistant Chief, appointed by the Secretary of the Navy and designated, respectively, the Chief and the Assistant Chief of Research and Inventions. In addition to reporting directly to the Secretary of the Navy, the Chief of Research and Inventions shall also report to the Chief of Naval Operations. The orders of the Chief of Research and Inventions shall be considered as emanating from the Secretary of the Navy and shall have full force and effect as such.

William Tompkins, *Selected by Extraterrestrials*, p. 315

interview, Tompkins made the startling claim that it was the Nordics who had recommended Admiral Botta to Forrestal as the person who should head the Navy espionage program in Nazi Occupied Europe.

> What's most important here, and frankly I would say that this is definitely not common knowledge. Secretary of the Navy [Forrestal] had a communication from Nordics to make a selection of the individual who was to take this [Navy espionage] program. That individual [Botta] was not even an American. He was from Australia. He went to school here. Joined the Navy, went up the ranks very fast, to finally a one star admiral. But what's important here is that of the 37 admirals that could have been picked to handle this, an individual who had not been influenced by having graduated from Annapolis, or gotten all of the incorrect information about everything on this planet, which is misinformation that's been given to us by the Reptilians for at least 5000

> years, controlling this planet. None of those Annapolis graduates were selected, but an individual was selected that had never gotten a university education. And I think this is extremely important, because virtually every Ph.D. on this planet has been given incorrect information. It is the major problem that we have in this subject as far as disclosure is concerned. Because the Ph.D's have been lied to, all of them! Why was it that Secretary of the Navy Forrestal, selected an Australian man, without an education to be the implementer of the most important information that's ever taken place on this planet. We got to ask questions, period.[100]

Botta's technical expertise and leadership skills during WWII were recognized by the Navy, which had rapidly promoted him, and bestowed various military honors as historical records prove. Forrestal's choice of Botta to lead the Naval intelligence program had been a good one.

If Tompkins is correct, Forrestal was directly working with Nordic extraterrestrials, who influenced him not only to choose the Naval officer who would establish the Navy's Research and Development of Nazi flying saucer technologies, but to also steer the US administration away from cooperating with the Nazis and Reptilians, and to reveal the truth to the public.

However, Forrestal was in a small minority among the policy makers dealing with the Nazi and extraterrestrial problem through an appointed committee set up by President Truman in 1947, which was called Operation Majestic 12 (see Truman Memo, Figure 30). The Majestic 12 Committee was not in favor of any kind of public disclosure of the alien problem.

The Air Force Secretary Symington was a firm supporter of General Curtis LeMay, who "considered space operations to be an extension of air operations," and therefore supported the US Air

Figure 30. Truman Memorandum that Authorized Operation Majestic 12

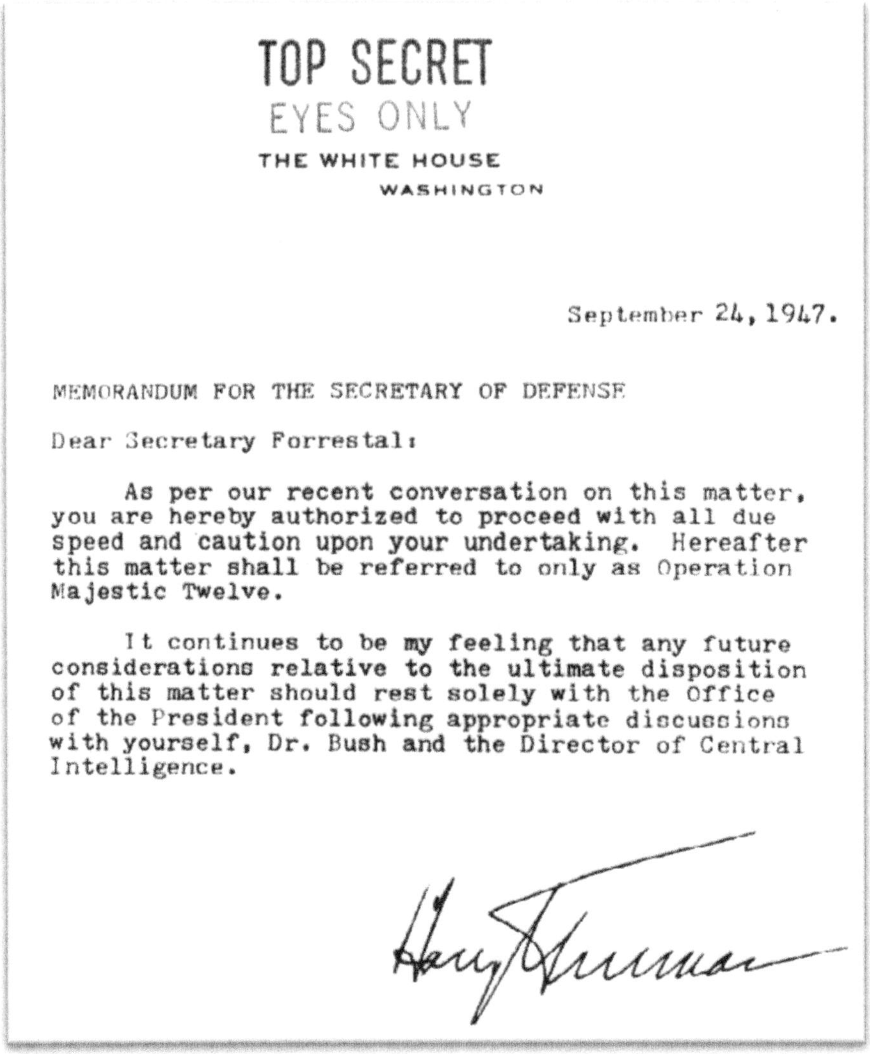

TOP SECRET
EYES ONLY
THE WHITE HOUSE
WASHINGTON

September 24, 1947.

MEMORANDUM FOR THE SECRETARY OF DEFENSE

Dear Secretary Forrestal:

As per our recent conversation on this matter, you are hereby authorized to proceed with all due speed and caution upon your undertaking. Hereafter this matter shall be referred to only as Operation Majestic Twelve.

It continues to be my feeling that any future considerations relative to the ultimate disposition of this matter should rest solely with the Office of the President following appropriate discussions with yourself, Dr. Bush and the Director of Central Intelligence.

Harry Truman

Force in a dominant role when it came to developing future space programs.[101] Both Symington and LeMay were firmly opposed to Forrestal's attempt to give the Navy the leadership role when it came to space, Nazi spacecraft, and extraterrestrial life. Furthermore, Symington and LeMay were vehemently opposed to revealing the truth of the extraterrestrial problem to the general public.

These policy disputes were a direct factor in the events that led to Forrestal's sacking as Secretary of Defense on March 28, 1949, and "assassination" two months later. Significantly, this was the same period when the Navy and Army Air Force collaboration in Project RAND, through the Douglas Aircraft Company, came to an end. The RAND Corporation was launched as an independent corporate entity in early 1948.

Figure 31. James Forrestal Receives Distinguished Service Medal from President Truman in March 1949

The second issue to consider is a proxy extraterrestrial war involving separate USAF and Navy led secret space programs. There are a number of whistleblower testimonies describing that the USAF led the way in a secret treaty with a group of extraterrestrials directly connected with the Nazi-Reptilian alliance. In February 1955, a meeting occurred at Holloman Air Force Base, where President Eisenhower secretly negotiated an agreement with a group of Nazi linked extraterrestrials, called the "Tall Grays."[102] As stated by a number of whistleblowers, the USAF worked directly with extraterrestrials and Nazis behind the 1955 agreement to develop squadrons of antigravity spacecraft, in the 1960's and 1970's, for a USAF run secret space program.[103]

Charles Hall is a former weather observer for the US Air Force from 1963 to 1967. He was stationed at the Nellis Air Force Base's Indian Springs facility, where he claims he frequently witnessed extraterrestrials called the "Tall Whites" (aka Tall Grays) regularly meeting with senior military leaders. A secret underground base was built at Indian Springs to house the extraterrestrials and their advanced interstellar ships. Rumors of the extraterrestrials at Nellis AFB date back to the mid-1950's, which is consistent with testimonial evidence that agreements were reached during the Eisenhower administration. Confirmation has been found for some of Hall's claims concerning anomalous events at the weather ranges, including government funding for a secret underground base at Indian Springs.[104]

Hall explained the legalistic way in which agreements with the Tall White extraterrestrials are interpreted, in this response to an interview question, about why he was chosen to liaise with them:

> ... the decision to send me, and no one else, out to the ranges, was made by a committee of individuals that included the Tall Whites as well as high ranking USAF Generals and other high ranking members of the U.S. Government. The Tall Whites are very meticulous about keeping their agreements and expect the U.S. Government to be equally meticulous about keeping its agreements as well. If I were victimized or threatened by anyone, The Tall Whites would interpret that to mean that the U.S. Government could not be trusted to keep its agreements. The consequences would be enormous.[105]

Hall's testimony is significant since it reveals that extraterrestrials are supplied a number of resources, including basing rights in exchange for technological assistance to the USAF. Secretly formalized by officials in the Pentagon and other key government agencies, one or more undisclosed agreements have arisen out of face-to-face meetings between both Air Force and national security officials with the Tall Whites.

Another former Air Force employee with direct first-hand experience of Gray extraterrestrials working with the USAF is Niara Isley. She worked for the US Air Force in the late 1970's as a radar specialist. During a three month period from January to March 1980, she was involuntarily recruited into a black project after being asked to get a radar lock on a UFO at the Tonopah Test range. After successfully performing her assigned mission, she describes what happened as a consequence of her viewing a UFO:

> I was dragged down what seemed like an abnormally long staircase through another door in the room. I was placed on the floor of a room with a one-way mirrored observation glass, mirrored of course on my side. Locked in there, I went through the effects of the injection, which was terrifying also. I can only describe feeling like I was coming apart at the molecular level. I don't remember pain, only the fear of dissolving away into nothing. After the effects of the injection were beginning to fade, I was dragged out of the room and raped by two security guards while eight other people watched, one of them a Grey extraterrestrial. I remember quite a bit of detail of this and can draw elements from these memories in detail.[106]

Isley's testimony is further evidence that Gray aliens actively collaborate with members of the US military in a secret space program that routinely violates human rights. The cooperation between the USAF and Grays, and other Reptilian allied extraterrestrials, extends to this day.

The RAND Corporation continues to help the USAF develop a coherent policy in terms of global research and development related to its secret space program, which is entirely separate to the very public NASA space program. RAND, then and now, in collaboration with the US Air Force Space Command, secretly studies antigravity designs that could help develop new squadrons of antigravity spacecraft in the future.

Currently, USAF Space Command uses antigravity craft developed with extraterrestrial assistance, such as the TR-3B flying triangle out of Area 51 in Nevada, for near-Earth space operations.[107] The flying triangles service two stealth space stations which secretly orbit the Earth. In response to a question about whether astronauts on the International Space Station observe the activities of secret space programs, Goode reveals critical information about the USAF covert space stations:

> Yes, they [ISS astronauts] absolutely do observe activity going on around them. For the most part, they're seeing the military, the MIC SSP, the program that is controlled by the NSA, the DIA, Air Force, that kind of thing. They have a couple of space stations up there that every time they orbit the Earth, they get a look at. So they're seeing these space stations that are about 400 miles to 500 miles out, and then they see the unacknowledged craft [TR-3B flying triangles] that are servicing these space stations.[108]

In contrast to the close USAF cooperation with RAND, the next chapter reveals that up to the early 1960's, the US Navy worked closely with Douglas Aircraft's Advanced Design in designing space battle groups with massive space carriers and cruisers. According to Tompkins, the Navy and Douglas were assisted by a very different group of extraterrestrials.

Endnotes

[69] Feb 22, 1944, Roosevelt Memorandum, Majestic Documents, http://majesticdocuments.com/pdf/fdr_22feb44.pdf
[70] Private telephone interview, September 19, 2016.
[71] William Tompkins, *Selected by Extraterrestrials*, p. xv.
[72] William Tompkins, *Selected by Extraterrestrials*, p. xii.
[73] Tompkins, *Selected by Extraterrestrials*, p.58.
[74] Phone interview with William Tompkins, September 19, 2016.
[75] Tompkins, *Selected by Extraterrestrials*, p.104.
[76] William Tompkins, *Selected by Extraterrestrials*, p. xv.
[77] William Tompkins, *Selected by Extraterrestrials*, p. 192.
[78] William Tompkins, *Selected by Extraterrestrials*, p. 192.
[79] "Twining's 'White Hot' Report," *The Majestic Documents* (Wood and Wood Enterprises, 1998) p. 75
[80] See Majestic Documents website, http://tinyurl.com/jt49ov3
[81] William Tompkins, *Selected by Extraterrestrials*, p. 192.
[82] Smith's memo is available online at: http://www.majesticdocuments.com/pdf/smithmemo-21nov51.pdf
[83] Wesley Marx, "The Military's 'Think Factories'," *The Progressive*, https://www.cia.gov/library/readingroom/docs/CIA-RDP88-01315R000400280026-3.pdf
[84] "A Brief History of RAND," http://www.rand.org/about/history/a-brief-history-of-rand.html
[85] Wesley Marx, "The Military's 'Think Factories'," *The Progressive*, https://www.cia.gov/library/readingroom/docs/CIA-RDP88-01315R000400280026-3.pdf See also Wikipedia, https://en.wikipedia.org/wiki/Henry_H._Arnold. Accessed 10/19/16. This figure is substantially more than what the RAND Corporation website describes. http://www.rand.org/about/history/a-brief-history-of-rand.html
[86] According to the website, "Measuring Worth," 10 million dollars in 1945 converts to between 107 to 790 million in 2016 dollars. https://www.measuringworth.com/uscompare/relativevalue.php
[87] "A Brief History of RAND," http://www.rand.org/about/history/a-brief-history-of-rand.html
[88] "Preliminary Design of an Experimental World-Circling Spaceship," http://www.rand.org/pubs/special_memoranda/SM11827.html

---

[89] "Preliminary Design of an Experimental World-Circling Spaceship." Santa Monica, CA: RAND Corporation, 1946. http://www.rand.org/pubs/special_memoranda/SM11827.html. Also available in print form.
[90] Tompkins, *Selected by Extraterrestrials*, p.427.
[91] For detailed analysis of the conflicting Army Air Force Press announcements, see Thomas Carey and Don Schmitt, *Witness to Roswell: Unmasking the Government's Biggest Cover-up* (New Page Books, 2009)
[92] "A Brief History of RAND," http://www.rand.org/about/history/a-brief-history-of-rand.html
[93] Private telephone interview, September 19, 2016.
[94] Tompkins, *Selected by Extraterrestrials*, p.58.
[95] Tompkins, *Selected by Extraterrestrials*, p.58.
[96] Tompkins, *Selected by Extraterrestrials*, p.58.
[97] This will be discussed in chapter four.
[98] Tompkins, *Selected by Extraterrestrials*, p.73.
[99] "The Death of James Forrestal," http://tinyurl.com/j5h78y2
[100] "William Tompkins Answers Some Viewer Questions." Jan 12, 2017. https://youtu.be/5J5Vl6wkMpM?t=49m15s
[101] "Preliminary Design of an Experimental World-Circling Spaceship," http://www.rand.org/pubs/special_memoranda/SM11827.html
[102] "Ike and UFO's," *Exopolitics Journal* 2:1 (2007): http://exopoliticsjournal.com/vol-2/vol-2-1-Exp-Ike.htm
[103] See Michael Salla, *Exposing US Government Policies on Extraterrestrial Life* (Exopolitics Institute, 2009)
[104] See Michael Salla, "Further Investigations of Charles Hall and Tall Whites at Nellis Air Force Base: The David Coote Interviews," http://exopolitics.org/Exo-Comment-36.htm
[105] "Charles Hall and the Tall Whites: Another perception of the extraterrestrial phenomenon and the Area 51," http://karmapolis.be/pipeline/interview_hall_uk.htm
[106] Niara Isley, "Nellis AFB Radar Specialist Witnesses UFO," http://www.ufodigest.com/news/1208/dreamland.html
[107] The TR-3B will be discussed in chapter 10. See also Michael Salla, "Tom DeLonge & UFO Disclosure: Rocking the Secret Space Programs Boat – Pt 2," http://exopolitics.org/tom-delonge-ufo-disclosure-rocking-the-secret-space-programs-boat-pt-2/
[108] Transcript of "Cosmic Disclosure: Viewer Questions Part 5," https://spherebeingalliance.com/blog/transcript-cosmic-disclosure-viewer-questions-part-5.html

CHAPTER 4

INSIDE THE DOUGLAS THINK TANK & ITS INFILTRATION BY NORDIC EXTRATERRESTRIALS

Over a 12 year period, beginning in 1951, William Tompkins worked for an above Top Secret think tank within the Douglas Aircraft Company designing antigravity spacecraft, covertly requested by the US Navy. Tompkins says he was given the job at "Advanced Design" due to the exceptional skills he exhibited in his wartime service with Naval Intelligence from 1942-1946.

It was throughout his service at San Diego's Naval Air Station that Tompkins directly participated in intelligence debriefings of the Navy agents working within Nazi Germany's most secret aerospace facilities during and immediately after World War II. Tompkins said that over his four years with the Naval intelligence program, he helped in the covert distribution of data from Nazi Germany's two distinct secret space programs to Douglas Aircraft Company, along with other select aerospace companies and universities that had the scientific expertise to understand what the Nazis were doing.

When Tompkins joined Douglas Aircraft Company in 1950, the Advanced Design think tank still existed to study and design antigravity spacecraft, although no longer as part of Project Rand (1945-1948). A Majestic document called the "White Hot Report" provides an independent source corroborating that Douglas was involved in highly classified studies of retrieved alien spacecraft through Project RAND.[109]

Designing Navy Space Battle Groups

Once Tompkins moved within Douglas over to Advanced Design in 1951, he was specifically tasked to design a variety of antigravity space vehicles, using his knowledge of the Naval Intelligence gathered on approximately 30 Nazi German prototypes, and his own talent for detailed technical designs. Tompkins describes the two individuals who were his superiors at the Advanced Design think tank:

> I reported directly to Dr. [Wolfgang] Klemperer and Elmer Wheaton, the V.P. of engineering who wore two hats. He was V.P. of all the classified missile and space-systems programs. Unknown to 99.9%, Wheaton was V.P. of the above top secret compartmentalized extraterrestrial threats research Think Tank, too, sometimes referred to as Advanced Design.[110]

Wheaton and Klemperer were leading experts in missile and space-systems, and co-wrote chapter 15 in the 1946 Project Rand study for the development of the world's first earth orbiting satellite, where they examined the time and cost of the project.[111] In addition to their known expertise in conventional rocketry, Tompkins has supplied a document confirming that Wheaton and Klemperer were also researching UFO reports and antigravity studies in open source material. The document confirms that Tompkins' immediate superiors in the Douglas think tank were indeed studying antigravity propulsion theories during the 1950's, and this would directly aid them in overseeing the designs of large spacecraft for the Navy.

Tompkins says that he approached his work by developing mission parameters for the requested future space battle groups. He then was able to come up with designs that would allow the Navy to fulfill its forecasted space missions.

Figure 32. Memorandum Confirming Douglas Study of Antigravity Propulsion

MTM-622, Part 2

March 1, 1955

To: E. P. Wheaton, A-250

From: W. B. Klemperer, A-250

Subject: UNCONVENTIONAL PROPULSION SCHEMES

Copies to: H. Aurand, A-250; R. Demoret; A-250; J. B. Edwards, A-250; S. Kleinhans, A-250; T. A. Kvaas, A-250; H. Luskin, A-250; C. C. Martin, A-215; G. M. Files

Reference: MTM-622, December 20, 1954 (Declassified)

Our studies of the possible merits or significance of occasionally appearing publications concerning Unconventional Propulsion Schemes have been casually continued since writing the first memorandum (MTM-622) about their progress to mid December 1954.

Between that time and the end of February 1955, twenty more papers on pertinent topics have been obtained and read. They are reviewed in the appended Astronautical Literature Review pages, serial 026 to 045. The content of most of them falls into similar categories as those reviewed before, under serial numbers 001 to 025.

Several more occasions were had to talk personally to people about the subject. Two of such interviews are abstracted, one with Dr. C. B. Millikan and the other with Captain W. T. Sperry of American Airlines who encountered an UFO in flight in 1950.

We have also looked at a few typical "Flying Saucer" books but found none of them of technical significance thus far. Brief reviews of six of them are appended. A print of a color film tracking two Unidentified Foreign Objects near Missoula, Montana, was obtained. It is now being analysed by Iconolog techniques.

Correspondence was exchanged with Aviation Studies (International) Limited, 20-31 Cheval Place, Kinghtsbridge, London SW 7, England, who describe themselves as Management Consultants and who prepare and distribute the Aviation Reports discussing technical, commercial and political developments in the world of aviation, as mentioned in paragraph 1. Reference was made by us particularly to the article "Gravitic Steps" in their issue No. 357 of 19 Nov. 1954 (p. 531) in which veiled intimations were made of promising experimental results with a test rig; specific questions concerning details of these alleged experiments were submitted to the editor of the British publication. An answer dated 4 February 1955 was promptly received. In this reply, signed by R. G. Worcester (Director of Aviation Studies (International) Limited) we were referred to "an unclassified report on Project Winterhaven

> Every possible mission was considered for spacecraft carriers, battle cruisers, and support star ships capable of missions to the southeast quadrant, spiral arm of our Milky Way Galaxy, and its stars and planets. By conceiving naval space missions utilizing Think Tank Naval attack destroyer and Navy/marine assault space ship carrier configurations ... unsolicited bids could be submitted to the Navy by Douglas.[112]

Tompkins describes in his autobiography the covert way in which the Navy went about making design requests to Advanced Design, and accepting unsolicited proposals:

> After receiving our unsolicited proposal for star ships [the Navy put out a sole source request for a proposal for exploratory star mission vehicles...]. Actually we didn't even get an RFP (Request for Proposal); it was just slipped in under the floor door to our Advanced Design.... on the envelope it only said: "To Whom it may concern." [113]

Admiral Bobby Ray Inman was allegedly an intermediary for the unsolicited Douglas bids for Navy contracts to design the interstellar craft. In his autobiography, Tompkins describes a conversation between himself and Elmer Wheaton about how to get a bid accepted by the Navy:

> "Don't worry about ONR's [Office of Naval Research] approval," Wheaton said: They can get it for us." Then he said, "Okay, great job, gentlemen. If we can get this through the heavies in the other part, we might get a final out. I think ONI [Office of Naval Investigations] and ONR will both run with it. This is the kind of backup Bobby Ray (Bobby Ray Inman) needs to convince Forrestal's people that we can close the gap with the other ones out there. I thought, who are they? And who are the heavies in the other part?[114]

Dr. Robert Wood, the book's editor, added the following clarifying note immediately after Tompkins' above quote:

> My interpretation of these remarks and their chronology is this: Elmer Wheaton had contact with the UFO-cleared group in the Navy, which he referred to as "Forrestal's people" as the ones who knew about the UFOs issues. One of the new young Navy officers who was cleared for the UFO topic appears to have been Bobby Ray Inman, and his inside knowledge of the UFO problem may well have been the special link to his subsequent highly successful career. Apparently Bobby Ray was the main person interacting with the Wheaton think tank at the time of this conversation. Since Bill Tompkins time in this vault spans several years [1951-1963], it is not real clear that this conversation occurred in 1952 or perhaps a year or so later. [115]

I was able to reach Admiral Inman and asked him about Tompkins' claim. He said: "There is no way in the 50's or 60's I could have had any knowledge or been involved at all with anything Douglas Aircraft was doing."[116] Regarding his firm denial of any involvement with the Douglas Company and its projects, it is important to point out that he would have been, and likely still is, officially obliged to deny any involvement if the project had been classified as an unacknowledged special access program.

According to Tompkins, the projects at Douglas received the highest security classification possible in the defense industry, which today would correspond to an Unacknowledged Special Access Program (SAP). A 1995 Department of Defense (DoD) Manual titled: "National Industrial Security Program Operating Manual" describes the security obligation of those participating in a SAP:

> There are two types of SAPS, acknowledged and unacknowledged. An acknowledged SAP is a program which

> may be openly recognized or known; however, specifics are classified within that SAP. The existence of an unacknowledged SAP or an unacknowledged portion of an acknowledged program, will not be made known to any person not authorized for this information.[117]

Despite his retirement, Inman would still be officially required not to disclose the existence of, or his involvement in, any unacknowledged SAPs involving the Navy and Douglas Aircraft Company. Inman's career postings do raise the possibility that he was a Navy liaison between the ONI and Douglas Aircraft during Tompkins' time in Advance Design up to 1963.

In my December 1, 2016 phone interview with Inman, he said that he began his Naval Commission in March 1951, and after several assignments, attended a Naval Intelligence postgraduate program in 1957 in Washington DC. He further stated that after graduation, he stayed on at the Pentagon as an intelligence briefer until 1960. Then after another short assignment, he was sent to a Navy office in the National Security Agency from October 1961 to 1965 as an Intelligence specialist. During his time at the Pentagon and/or NSA, he could have played a liaison role between the Navy and Douglas Aircraft's Advanced Design, as Tompkins claims.

Tompkins also said that Advanced Design worked closely with a number of Navy research development facilities such as Naval Development Center, Warminster, Pennsylvania; Naval Air Weapons Station, China Lake; and the Naval Air Station, San Diego.[118] He asserts that 75% of the technical projects at Advanced Design were funded by the Navy, and admirals could often be found at the think tank. In contrast, it was rare to see an USAF general at Douglas Aircraft's Advanced Design. The USAF focus had changed and it became the major funder of the RAND Corporation after it separated in 1948 from the Douglas Aircraft Company.

How he went about creating the configuration of a Naval Space Battle Group, comprising kilometers-long vehicles from the mission parameters he had been given, Tompkins explains:

> I redefined a standard Naval space battle group complement, stating that it would consist of one 2.5 kilometer spacecraft carrier, with a two-star on board as flag, three to four 1.4 k heavy space cruisers, four to five 1k space destroyers, two 2k space landing assault ships for drop missions, two 2k space logistic support ships, and two 2k space personal transports.[119]

Tompkins writes about two Navy starship designs completed at the Douglas think tank, and includes the documents in his autobiography:

> The figures following show two original drawings of naval spacecraft carriers and battle cruisers that were visualized in Advanced Design, in 1954, from dozens of alternate configurations. Scale modes of these kilometer-long craft were subsequently made.[120]

In one of the two designs, Tompkins used for comparative purposes the USS Akron, a Navy airship deployed in 1931 that was 240 meters (785 feet) in length. It could carry "F9C Sparrowhawk" and "Waco XJW-1" aircraft, and was the world's first flying aircraft carrier.

The USS Akron, and its sister ship, the USS Macon, were built as a joint venture between the US based Goodyear and German based Zeppelin companies. The Akron class airships confirm that the US Navy was already involved in building flying aircraft carriers in the late 1920's. The Navy, therefore, had the history and expertise to begin work on developing similar space carriers once antigravity propulsion technologies became feasible in the 1950's.

Figure 33. Sparrowhawk Fighter being pulled into USS Akron

Figure 34. USS Akron: helium-filled rigid flying aircraft carrier that crashed in April 1933

In Tompkins' design for a heavy transport spacecraft that was three kilometers (two miles) long, he compared it in size to the USS Akron. He showed how it would be built using a modular construction process with nearly 15 million '10 foot' cubes.

Figure 35. Massive Space Carrier

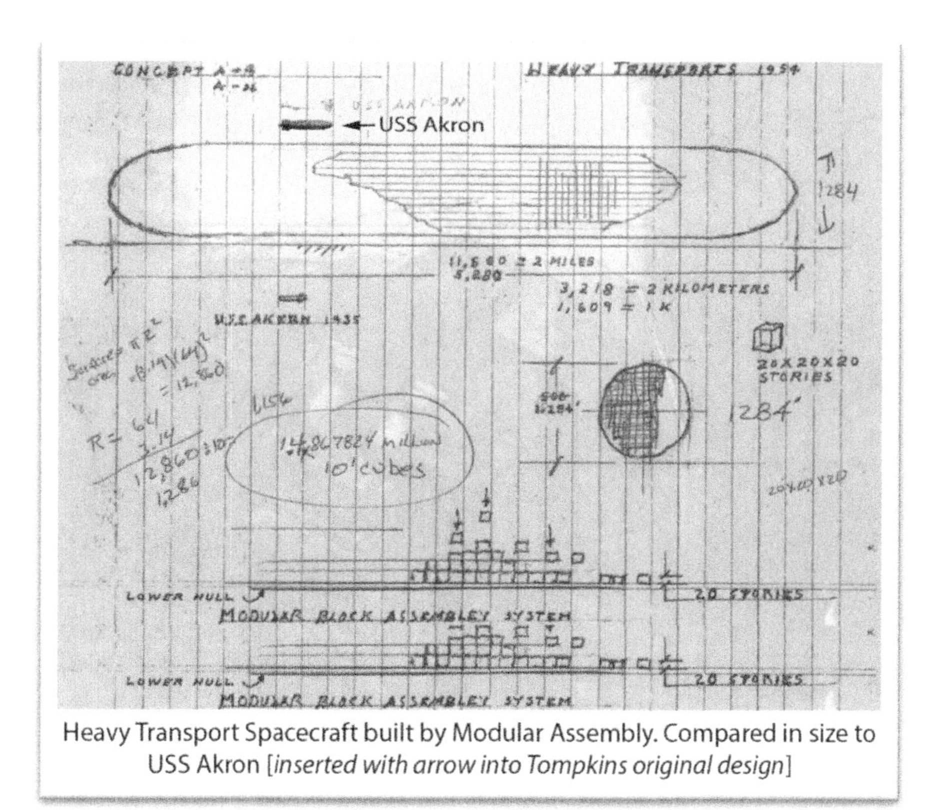

Heavy Transport Spacecraft built by Modular Assembly. Compared in size to USS Akron [*inserted with arrow into Tompkins original design*]

Tompkins describes how by 1952, just two years after he had joined Advanced Design at Douglas, he and his colleagues had already accumulated a large number of designs for the different classes of vehicles for the future Navy space battle groups:

> At the time, we in the tank, had already conceived a file cabinet full of Naval Space Missions and another full of Naval spacecraft ships from spacecraft carriers to spacecraft transports, six years before NASA even existed [created on July 29, 1958]. A great number of these missions became NASA's (NAVY) prototype for our next thirty year penetrations to solar system planets and stars.[121]

Tompkins provided another document showing a mile long (1.6 kilometer) battle group flagship. In the diagram, the annotation shows that the design was originally conceived at Douglas'

Figure 36. Space Battle Group Flagship

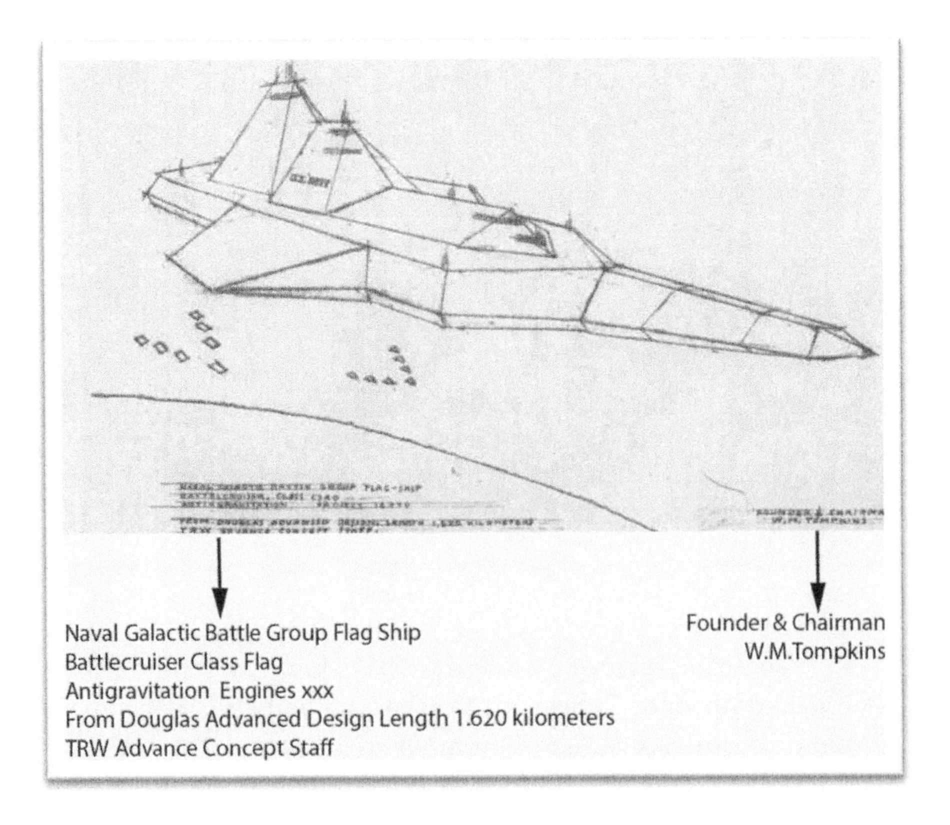

Naval Galactic Battle Group Flag Ship
Battlecruiser Class Flag
Antigravitation Engines xxx
From Douglas Advanced Design Length 1.620 kilometers
TRW Advance Concept Staff
Founder & Chairman
W.M.Tompkins

"Advanced Design," and re-introduced at a similar think tank within TRW, called "Advanced Concept," where Tompkins worked from 1967 to 1971.

The next graphic displays a close up of the hull of the flagship showing how it would operate as a space carrier for smaller triangular shaped fighter spacecraft (similar to the TR-3B to be discussed in chapter ten).

Figure 37. Triangle Spacecraft Entering Flagship

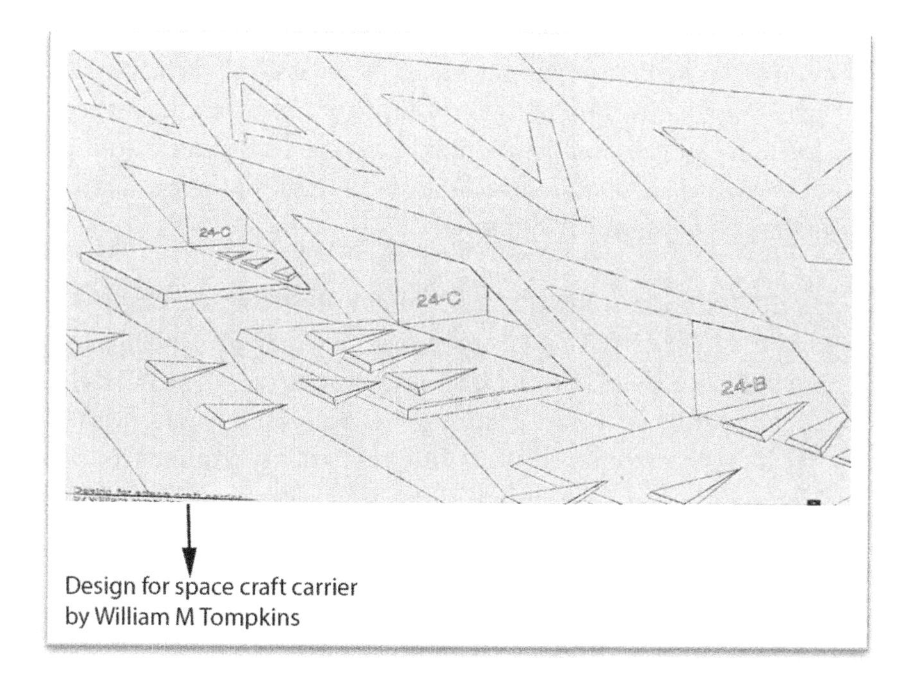

In an interview on Gaia TV, Tompkins explained:

> And what you see there is the different classes of attack and fighter aircraft returning to the mothership, or the spacecraft carrier, and with vacuum-controlled entry sections . . . Actually, they would design to fold down so that you're already in support of making inside landing.

> Nobody actually flies in these squadrons controlling any of the vehicles. It's all automatic, so you're not going to be hitting the walls or any of this kind of thing. But very quickly, you can open these hatches on the side. The hatch then becomes sort of a platform to possibly land on if you're too low. And this answers the question of how do we handle the large spaces aboard the ships that are going to be operating on the missions.[122]

While the designs for the enormous space battle groups were being conducted at Douglas, and later at TRW, the detailed engineering and construction occurred at a classified facility in Utah near the Wasatch Mountains, according to Tompkins. In response to a question on this subject, Tompkins told of a conversation with "Admiral Roscoe" (Hillenkoetter) about completing the project, and wrote about it in his autobiography:

> Well Admiral ... as you know, we here on the working side of the DAC/RAND Think Tank – are proposing to build a large spaceship development and manufacturing plant in Utah. It can handle your two-kilometer spacecraft carriers, with the capability to expand the facilities to accommodate ships that are ten times that size. We'll need the funding, of course. That's where you come in, Admiral.[123]

In a radio interview, Tompkins provided more details about the construction facilities for the kilometers-long spacecraft that was designed at Douglas, and the aerospace companies involved in the construction process:

> One of the facilities that was used is east of the Wasatch Mountains in Utah. This was a very, very large cavern. It has smaller ones adjacent to it, but it's a massive one. And they were able to put together Lockheed space systems, Northrop-Grumman facilities, and even Boeing were involved in the actual construction of these kilometer-long

spacecraft carriers that the Navy now has eight of these spacecraft carrier battle groups out in the galaxy.[124]

It is worth pointing out that Lockheed Martin (formally Lockheed) and Northrup Grumman (formally Northrup), along with Douglas Aircraft Company, had received briefing packets from Tompkins from 1943 to 1946. Tompkins says that Lockheed's top secret think tank, "Skunk Works," which was officially formed in June 1943, emerged as a direct result of the contents of the briefing packets he delivered.[125] In 1997, Boeing absorbed McDonnell Douglas, which had previously formed out of the 1967 merger between Douglas Aircraft Company and McDonnell Aircraft. Consequently, the three main corporations responsible for building and upgrading the Navy's eight space battle fleets, which were constructed in secret Utah locations since the 1980's, can all be directly traced to the original companies that received Tompkins' briefing packets.

Nordic Extraterrestrials Infiltrated Douglas Aircraft Company

Tompkins says he and his colleagues were helped by three Nordic extraterrestrials secretly posing as normal citizens, who had gained employment within the Douglas Aviation Company. He described how these three Nordics, two females and one male, provided key information about the design and construction of space vehicles without revealing their true identity. Tompkins even explained to a co-worker in Advanced Design how his Secretary, Jessica, would not reveal her true origin, but was clearly one of the Nordic extraterrestrials assisting him in completing Douglas company projects on behalf of the Navy:

> I tried to explain to John that Jessica has never said that she is a Nordic, even though ... exposing her unbelievable knowledge of the universes. And yes, she frequently stuffs my head with exactly what is needed on a program. When I implement the plan it works every time. [126]

Upon seeing a photo of Maria Orsic, which Tompkins saw for the first time on page 67 of the book *Insiders Reveal Secret Space Programs and Extraterrestrial Alliances* (2015), he was startled. This is because she looked exactly as he remembered Jessica, the Nordic extraterrestrial who had infiltrated Douglas Aircraft Company as a secretary.[127]

Figure 38. Tompkins said Jessica (a Nordic ET) looked exactly like Maria Orsic

Tompkins' claim, that the Douglas Aircraft Company has been infiltrated by three Nordic extraterrestrials, is supported by multiple whistleblowers and private citizens that claim human-looking extraterrestrials have successfully integrated into all facets of human society, including the most senior levels of the world's military, including NATO.[128]

Robert Dean (CSM, US Army Ret.) worked at NATO's Supreme Headquarters from 1963-1967, and during this time was stationed in the Operations Center with a Cosmic Top Secret clearance. He says he viewed a secret NATO study that was commissioned to analyze the threat posed by UFOs to NATO operations in Eastern Europe. The classified report was titled: "An Assessment: An Evaluation of a Possible Military threat to Allied Forces in Europe." It focused on the dangers of UFOs being mistakenly identified as an incoming ballistic missile attack from the Soviet Union.

Dean claimed that the NATO study identified four different extraterrestrial civilizations visiting the Earth at that time. According to his testimony, what really worried the NATO top brass was that some of the visitors looked so much like us that they were virtually indistinguishable. Dean says that NATO generals were paranoid over the possibility that some of the extraterrestrial visitors could be walking in the corridors of NATO or the Pentagon, or even the White House itself. In an interview Dean stated:

> There was a human group that looked so much like us that that really drove the admirals and the generals crazy because they determined that these people, and they had seen them repeatedly, they had had contact with them.... These people looked so much like us they could sit next to you on a plane or in a restaurant and you'd never know the difference. And being military and being primarily paranoid, that bothered the generals and the admirals a little bit. That the fact that these intelligent entities could be involved with us, walking up and down the corridors of SHAPE, walking

> down the corridors of the Pentagon. My God, it even dawned on a couple of them that these guys could even be in the White House! Of course, as I said, being paranoid in those years it really shook things up a little bit.[129]

Another public official to have disclosed the truth about extraterrestrials infiltrating and living among humans is the current Russian Prime Minister and former President, Dmitry Medvedev. On December 7, 2012, he made some startling off-air comments to reporters while his microphone was still switched on. He was asked whether the President is given any secret files on extraterrestrials while in office. In his responses, Medvedev not only confided that extraterrestrials are visiting the Earth, but that some are actually living among us:

> Along with the briefcase with nuclear codes, the president of the country is given a special 'top secret' folder. This folder in its entirety contains information about aliens who visited our planet... Along with this, you are given a report of the absolutely secret special service that exercises control over aliens on the territory of our country...[130]

Dean and Medvedev's statements are a vital key in unlocking the truth of extraterrestrials living among the human population, and infiltrating the military-industrial complex. His testimony conclusively demonstrates that official military and government agencies are aware of this penetration, and in fact have developed strategies for such a contingency.

Corey Goode says that for several years, he was employed in an intercept and interrogate program involving extraterrestrials found on Earth.[131] The intercept program would identify aliens that had infiltrated human society and bring them in for interrogation. Goode said that his task was to find their purpose for being on Earth, and detect any deception on the part of the extraterrestrial during their interrogation. Coercive methods such as torture were used on the alien infiltrators. What Goode witnessed was deeply traumatic, and he says he needed assistance in dealing with the trauma of the incidents which he still fully recalls. He said in an email to me:

> Their [Mayan] technology was all neurologically interfaced technology and was very impressive. The device they used to assist me in the removing of the "Pain" and "Negative Energy Associations" with certain very dark memories was interesting. It was what they called a "Halo" and looked like it was made of "Gold" but was light as a feather. When they put it on top of my head it "Sucked" to my skull/scalp like my head was a magnet. And they looked at a floating console and they never touched but interacted with mentally.... They were of great service to me and I am told they have been brought in to assist many of the Humans who have been "Recovered" from Bases and even from slavery in other solar systems. There are secret locations in other Sol Systems that have colonies for these humans to recover as they will not be able to reintegrate with our society anytime soon. This group has assisted them greatly. Many assume that they are "Aliens" that are assisting former Human Slaves. These people just do not have the full facts.[132]

In the case of the Nordics that had established their presence inside Douglas, Tompkins claims they assisted him greatly. In fact, the assistance he was receiving from his Nordic friends had elevated him into a preeminent position as an authority on different types of extraterrestrials and their technologies:

> Over time, both in and out of the Advanced Design Think Tank, I got the reputation as the principal thinker addressing the black hat alien [Draco Reptilians Empire] threats and conceiving programs to counter them. Somehow, other

> white hat aliens [Nordics] dropped concepts in between my ears that always worked. It was like I had some of their technology that was thousands of years more advanced than ours here on planet Earth.[133]

In a later chapter, both documentary and witness evidence will be introduced to show that Tompkins was indeed considered to be an authority on extraterrestrial life and technology by the retired Navy officers who worked with him on "Special Projects" with the Navy League in the 1990's.

Navy Approved Tompkins' Relationship with Nordic Extraterrestrials

Tompkins specifically recalls a conversation with Elmer Wheaton, who told him that the Navy considered him to have been chosen by the Nordics at a young age to play a critical role in being an intermediary between the Navy and Nordics:

> We know that several of you guys, especially you, have been selected by some of these alien guys, if you get my meaning. Naval intelligence wasn't alone when they took an interest in you. There are aliens – probably Nordics – who selected you as a child to assist them in their endeavors to prod the US into developing galactic Naval spaceships and operations advantageous to us both.... Our Naval Intelligence considers you a preferred human contactee. The communication link between these extraterrestrial races and you contactees provides us with advanced counter threat concepts, not just for the Apollo program, or for the Navy base on the Moon, but literally for all advanced space concepts for subsequent encounters....[134]

Tompkins recollections here are staggering in their implications. They suggest that key figures in the Office of Naval Intelligence were aware that Tompkins and others, such as Navy Secretary James Forrestal and Rear Admiral Rico Botta, had a special connection to a friendly human-looking group of extraterrestrials, which were in conflict with the same Reptilians who had formed an alliance with Hitler. This meant that these specially selected individuals, or "preferred human contactees," would form a critical role in helping establish a covert alliance between the US Navy and Nordic extraterrestrials.

Such an alliance had to be covert, due to the infiltration of the military industrial complex by Nazi Breakaway groups and their Reptilian allies, which had occurred as a result of formal agreements between the Nazi-Reptilians and the Eisenhower administration, beginning at Holloman Air Force Base in February 1955. The Nazi-Reptilian alliance would increasingly coopt many elements of the US military industrial complex over time, including key components of the Air Force, the RAND Corporation, the Central Intelligence Agency, and even the Douglas Aircraft Company.

In contrast, senior Navy officials covertly worked with the Nordics both to prevent Nazi-Reptilian infiltration, and to build future Navy space battle groups. In a Gaia TV interview with David Wilcock, Goode corroborated Tompkins' testimony of the Navy permitting Nordic extraterrestrials to infiltrate major corporations like Douglas Aircraft Company:

> David: How would one of these Nordic men or women be able to get into such a highly compartmentalized, highly classified thing? Wouldn't our government or the military-industrial complex be highly afraid of them because of the possibility of them either deceiving us or bringing intel back to their own group and then weaponizing it against us?

> Corey: Yeah, and after a while, they found out that we were being deceived about some things that the Nordics – that we're calling them – did for operational security, you know? There were definite[ly] some deceptions that were given to us. But for the most part, they were learning about the reptilian threat and the threat of their allies. And this group [Nordics] look a lot like us. Basically, it's like a cousin coming to help a cousin.
>
> David: The enemy of my enemy is my friend.
>
> Corey: Yeah, yeah. But these people presented themselves as wanting us to get rid of our nukes, wanting us to become more loving and peaceful. And they just . . . They [US Military] did not see as big of a threat from them [Nordics] as they did their enemy, the Reptilian groups.[135]

The fruit of this covert cooperation was the development in the 1980's of joint operations between the first Navy space battle groups and the Nordic extraterrestrial space fleets, which Tompkins describes here:

> I don't know the actual mission right now, but for most of the periods the first ones were built [1980's] and went out into the galaxy because they were built here. They were operating with only one of the eight who operated around the solar system. The others were out in the galaxy jointly operating missions against Reptilians and other bad people, but jointly working with the Nordic Navy, and who were the people that had been assisting at least us on the Apollo.[136]

Tompkins went on to illustrate the extent of the cooperation between the Nordic fleets and the Navy's Solar Warden program, and the disparity in numbers with the Reptilian imperial fleets:

> [W]e're operating, and have been, since the '80s, with the Nordic Navy. And these people . . . It's easiest to describe the situation there . . . For every ten Nordic Navy battle groups, the Reptilians have 100. So we're sort of at a disadvantage, by numbers, but a Nordic Navy has been tremendously supportive of us. Our Navy works with them and actually is at war not just with Reptilians out in space but several other bad civilizations.[137]

While the above scenario may sound like an episode of *Star Trek*, involving a galactic conflict between the fictional United Federation of Planets and the Klingon Empire, there is reason to believe that Gene Rodenberry was made aware of key elements of the truth. Rodenberry, through the son of a Navy admiral, was confidentially told about plans to build a Space Navy, and of a galactic war between Nordic extraterrestrials and the Reptilian Empire. Humanity found itself in the midst of this galactic wide conflict, and *Star Trek* was inspired in part to reveal key elements of this conflict in order to prepare humanity for the truth through a "soft disclosure" process.

Endnotes

[109] "Twining's 'White Hot' Report," *The Majestic Documents* (Wood and Wood Enterprises, 1998) p. 75
[110] William Tompkins, *Selected by Extraterrestrials*, p. 48.
[111] *Preliminary Design of an Experimental World-Circling Spaceship*. Santa Monica, CA: RAND Corporation, 1946. http://www.rand.org/pubs/special_memoranda/SM11827.html. Also available in print form.
[112] William Tompkins *Selected by Extraterrestrials,* p. 105.
[113] William Tompkins *Selected by Extraterrestrials,* p. 68.
[114] See William Tompkins *Selected by Extraterrestrials,* p. 12.
[115] See William Tompkins *Selected by Extraterrestrials,* p. 12.
[116] Phone interview with Admiral Bobby Ray Inman on December 1, 2016.
[117] "National Industrial Security Program Operating Manual:" DoD 5220.22-M-Sup. 1, February 1995. 1-1-2. Available online at: https://www.fas.org/sgp/library/nispom_sup.pdf
[118] Phone interview with William Tompkins, September 19, 2016.
[119] William Tompkins *Selected by Extraterrestrials,* p. 80.
[120] William Tompkins *Selected by Extraterrestrials,* p. 67.
[121] William Tompkins *Selected by Extraterrestrials,* p. 396.
[122] "Cosmic Disclosure: Founders of Solar Warden with William Tompkins, Season 6, Episode 8," http://spherebeingalliance.com/blog/transcript-cosmic-disclosure-founders-of-solar-warden-with-william-tompkins.html
[123] William Tompkins *Selected by Extraterrestrials,* p. 99.
[124] Rense Radio Interview with William Tompkins with Maj. George Filer & Frank Chille – May 4, 2016 http://spherebeingalliance.com/blog/the-amazing-story-continues-part1.html
[125] "Robert Wood and William Tompkins Interview - Part 3," Interview with William Tompkins, https://youtu.be/eblJYXe5iY8?t=57m20s
[126] William Tompkins *Selected by Extraterrestrials,* p. 274.
[127] Private interview January 16, 2016.
[128] See Michael Salla, "Extraterrestrials Among Us," *Exopolitics Journal*, vol 1:4 (2006). Available online at: http://www.exopoliticsjournal.com/vol-1/1-4-Salla.htm
[129] Bob Hieronimus, "Transcript of Interview with Bob Dean, March 24, 1996," published online at: http://tinyurl.com/jaxr6ef

---

[130] See Michael Salla, "Russian Prime Minister claims extraterrestrials live among us," http://exopolitics.org/russian-prime-minister-claims-extraterrestrials-live-among-us/

[131] He discusses his intercept and interrogate program in an interview, "Cosmic Disclosure: Contact Is Made," here, http://spherebeingalliance.com/blog/transcript-cosmic-disclosure-contact-is-made.html

[132] Email received on April 13, 2015, published online at: http://exopolitics.org/ancient-space-programs-human-extraterrestrial-alliance-meetings/

[133] William Tompkins *Selected by Extraterrestrials,* p. 182.

[134] William Tompkins *Selected by Extraterrestrials,* pp. 310-11.

[135] "Transcript, Cosmic Disclosure: SSP Think Tank with William Tompkins," http://spherebeingalliance.com/blog/transcript-cosmic-disclosure-ssp-think-tank-with-william-tompkins.html

[136] Rense Radio Interview with William Tompkins with Maj. George Filer & Frank Chille – May 4, 2016. http://spherebeingalliance.com/blog/the-amazing-story-continues-part1.html

[137] Rense Radio Interview with William Tompkins with Maj. George Filer & Frank Chille – May 4, 2016. http://spherebeingalliance.com/blog/the-amazing-story-continues-part1.html

CHAPTER 5

STAR TREK BASED ON SECRET NAVY SPACE FLEET

September 8, 2016 marked the 50th anniversary of the release of the "Star Trek" science fiction series that began in 1966 on NBC television. Its official creator, Gene Roddenberry, is highly respected for this series' groundbreaking concepts, yet there is significant evidence that he did not simply come up with the idea of *Star Trek* on his own. Instead, he was encouraged to create the series based on classified information surrounding the development of a secret US Navy space fleet that was allied with a group of human-looking extraterrestrials. This laid the foundation for Roddenberry to conceive the idea of a United Federation of Planets in which many worlds cooperated peacefully, sometimes sharing technology and personnel for space exploration.

Roddenberry began developing ideas for a science fiction show after one of his series had bombed in 1964:

> The only reason Roddenberry created Star Trek, at least initially, was to sell another series to a network. He was, if not desperate, anxious... He had just failed with The Lieutenant, for Norman Felton's Arena Productions ... No one was clamoring for another series from Roddenberry, or even his scripts. His agent suggested he come up with a space series... This may have led to what The Outer Limits

> historians insist are the accurate — if generally unknown — accounts of Roddenberry hanging out at times on the set of The Outer Limits. When I learned this, it wasn't hard to imagine that series creator, and executive producer, Leslie Stevens ... was someone that Roddenberry may have sought to emulate.[138]

This account is confirmed by Tom Seldon, one of the production assistants of the television series "The Outer Limits," which ran for two seasons (1963-1965) on ABC:

> Star Trek was in fact an outgrowth of The Outer Limits. Gene Roddenberry watched our dailies all the time and took a lot of phone calls from our screening room. He was spurring his imagination and checking on the incredible quality control we had. I wondered why he was there but he was there more often than not during the time he was coming up with Star Trek.[139]

Chris Knowles, a media blogger who wrote an eight part series of articles about the genesis of *Star Trek,* states that Roddenberry and Leslie Stevens IV had reached a business arrangement for the planned sci-fi series:

> Bearing in mind that Roddenberry was contracted to a rival studio and a rival network, the odds are essentially slim to none that the two men didn't have some kind of business arrangement, whether in writing or not.[140]

Gordon White, a book author, reaches the same conclusion about Stevens and Rodenberry's highly unusual relationship:

> The case is convincingly made [by Knowles] that Leslie Stevens - rather than Roddenberry - was at the terrestrial epicentre of Star Trek's prescient high strangeness, which it inherited from *The Outer Limits.* Having worked in broadcast

> production before, there really isn't any such thing as innocently sharing production resources. If Roddenberry was camped out with *The Outer Limits* team for a whole year, then that very much means something.[141]

What lends further credence to the Stevens and Rodenberry connection is that they both shared the same production assistant, Robert Justman, who worked on both the *Outer Limits* and *Star Trek*. According to White:

> For decades, Stevens and Roddenberry shared the same assistant, Rob Justman, passing him backward and forwards whenever one of them had a gig on. Justman would go on to be part of a military space futures experiment in the early nineties. [142]

The evidence firmly points to Stevens and Roddenberry having indeed developed a business arrangement for the new *Star Trek* series. This is where Stevens' background becomes critical in understanding the nature of their arrangement.

Stevens was born in 1924, and when his father, a US Navy officer, was assigned to the US Embassy in London in 1935, the following happened:

> ... the 11-year-old boy attended performances of Shakespeare at the Old Vic as part of his schoolwork and decided to become a playwright. Back in the States four years later he sold a play, The Mechanical Rat, to Orson Welles's Mercury Theatre and ran away from home to join the troupe. Truant officers tracked him down and he finished schooling with a Bachelor of Arts degree. [143]

In 1942, at age 18, he joined the Army Air Force and became an intelligence officer. Stevens quickly distinguished himself due to the skills he had acquired as a playwright, and ended the war as the youngest serving captain in the Army.[144]

As an intelligence officer, his playwriting skills would have suited him to psychological warfare operations. Consequently, it is possible that during his military intelligence service, Stevens learned about the Interplanetary Phenomenon Unit, set up by Army G2 to investigate the flying saucer phenomenon. Psychological warfare was a significant component of the Unit's work in "hiding the truth" about the phenomenon from the general public while seriously investigating it.[145]

Even more significant is the fact that Stevens' father was Navy Vice Admiral Leslie Stevens III. Vice Admiral Stevens was a contemporary of Rear Admiral Rico Botta, who had played a pivotal role in the Navy's initial efforts to study the retrieved flying saucer craft from the Los Angeles Air Raid, and oversaw a covert Navy espionage program out of Nazi Germany to learn about Nazi flying saucers during World War II. The 29 Navy spies in the program had

Figure 39. Leslie Stevens III & IV

Vice Admiral Leslie Stevens, III

Leslie Stevens, IV.
Creator of the Outer Limits

not only learned that the Nazis had developed up to 30 different flying saucer prototypes, but were also being directly assisted by an extraterrestrial civilization comprising Reptilian hominoids in building bases in Antarctica for future space battle fleets.

Both Admirals Botta and Stevens were leading experts in aerospace engineering and headed top Navy aerospace facilities at various points during their careers. Botta headed the Navy's Bureau of Aeronautics Power Plant Branch when he traveled to Wright Field and other top experimental aircraft facilities in 1942.

In 1946, both were retroactively promoted to the rank of Rear Admiral as of 1943, for their wartime duties as the following document shows. Botta's promotion was retroactively dated a

Figure 40. US Senate Approval of Admirals Stevens & Botta's Retroactive Promotions

EXECUTIVE CALENDAR

Saturday, May 25, 1946

NOMINATIONS

)ate of report	Calendar No.	Message No.	Name of nominee	Office	Predecessor
			IN THE NAVY APPOINTMENTS IN THE NAVY FOR TEMPORARY SERVICE		
24	849	237	John H. Towers	Admiral, to rank from Nov. 7, 1945.	
24	849	237	DeWitt C. Ramsey	Admiral, to rank from Dec. 28, 1945.	
24	849	237	Arthur W. Radford	Vice admiral, to rank from Dec. 28, 1945.	
24	849	237	Forrest P. Sherman	do.	
24	849	237	Lawrence B. Richardson	Rear admiral, to rank from Apr. 6, 1943.	
24	849	237	Rico Botta	Rear admiral, to rank from June 30, 1943.	
24	849	237	Leslie C. Stevens	Rear admiral, to rank from July 3, 1943.	

US Senate Approves Retroactive Promotion of Admirals Rico Botta & Leslie Stevens

Source: http://www.senate.gov/legislative/LIS/executive calendar/1946/05 25 1946.pdf

week before Stevens, suggesting he was the more senior when it came to their respective wartime command appointments. Botta went on to head the Naval Air Material Center out of Philadelphia Shipyard from 1950 to 1952.[146] It is from this and similar command assignments that plans began for a secret Navy space fleet aimed to counter what the Nazis had developed in Antarctica. Similarly, Admiral Stevens was known for his accomplishments in aeronautics, and is quoted as having "had a hand in the design or conception of all naval aircraft, aircraft carriers and carrier landing apparatus."[147]

Admiral Stevens' aeronautics expertise meant that he was almost certainly aware of what Botta had learned about Nazi aerospace projects beginning in early 1942, after the Los Angeles Air Raid incident. According to William Tompkins, Admiral Stevens was indeed made aware of what the Navy had learned about Nazi Germany's flying saucer programs.[148]

Documents reveal that Admirals Botta and Stevens did serve together on a line selection board for Navy Commanders. One document is dated January 6, 1950, and requests Admiral Botta to the Bureau of Naval Personnel to meet with a number of other admirals, one of whom was Stevens (see Figure 41). The document confirms that Botta and Stevens did meet in the fulfillment of their normal roster of duties, giving them the opportunity to discuss policy and technical issues related to the development of the Navy's secret space program. What is important here to keep in mind is that around the time Botta and Stevens served on this Board, Stevens had been assigned responsibility within the Joint Chiefs of Staff for clandestine activities and covert operations, as a declassified CIA document confirms (see Figure 42).

Soon after, Stevens was appointed to be the head of "The Joint Subsidiary Plans Division," which was formed in late 1949 under the control of the Joint Chiefs of Staff. Army historian Alfred H. Paddock describes the mission of this Division as follows:

Figure 41. Document shows Admirals Botta & Stevens serving on same committee

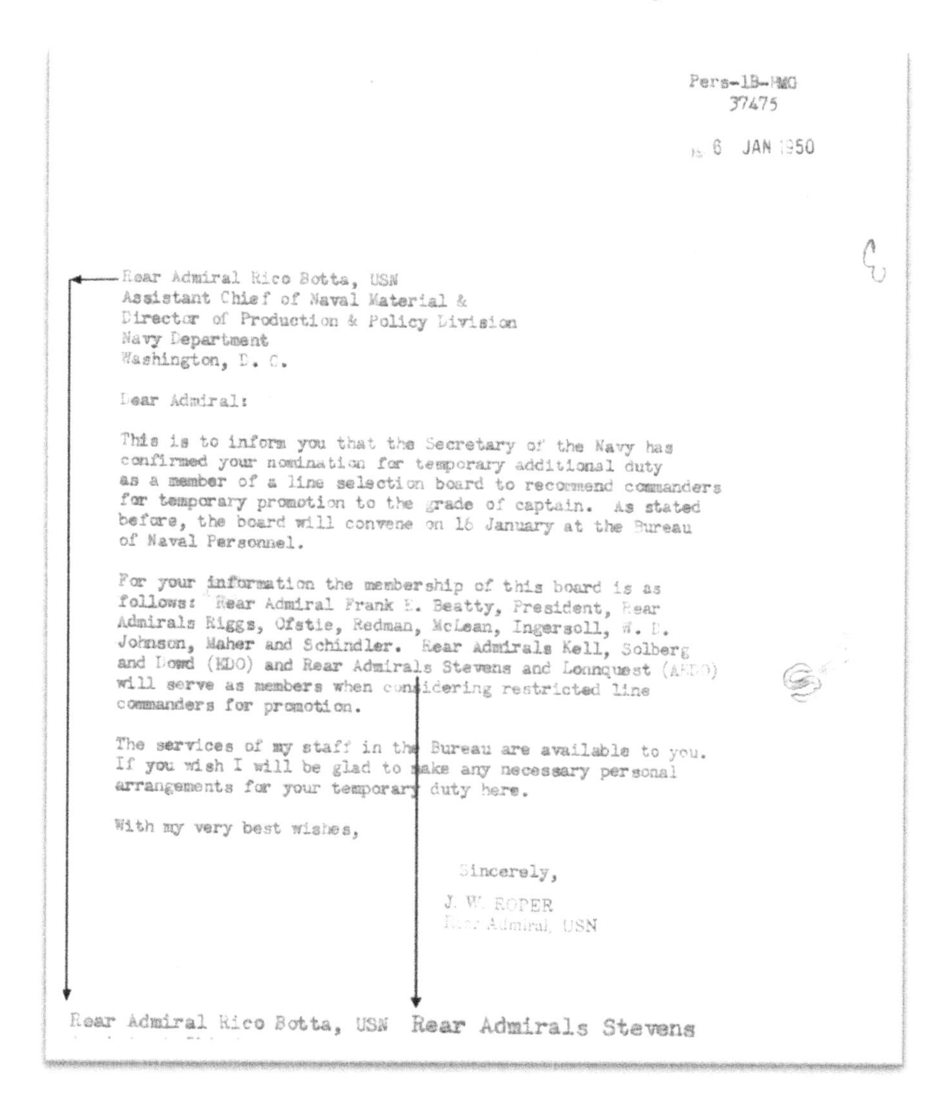

Pers-1B-HMG
37475

6 JAN 1950

Rear Admiral Rico Botta, USN
Assistant Chief of Naval Material &
Director of Production & Policy Division
Navy Department
Washington, D. C.

Dear Admiral:

This is to inform you that the Secretary of the Navy has confirmed your nomination for temporary additional duty as a member of a line selection board to recommend commanders for temporary promotion to the grade of captain. As stated before, the board will convene on 16 January at the Bureau of Naval Personnel.

For your information the membership of this board is as follows: Rear Admiral Frank E. Beatty, President, Rear Admirals Riggs, Ofstie, Redman, McLean, Ingersoll, W. D. Johnson, Maher and Schindler. Rear Admirals Kell, Solberg and Dowd (EDO) and Rear Admirals Stevens and Lonnquest (AEDO) will serve as members when considering restricted line commanders for promotion.

The services of my staff in the Bureau are available to you. If you wish I will be glad to make any necessary personal arrangements for your temporary duty here.

With my very best wishes,

Sincerely,

J. W. ROPER
Rear Admiral, USN

[To] coordinate the peacetime development of psychological warfare and covert operations capabilities within the Armed Services, coordinate detailed military plans and other agencies of the government, particularly with Department of State and the Office of Policy

> Coordination [CIA), and, in wartime, [to] become the means by which the JCS would provide continuous direction and guidance in these specialized fields to commanders under their control." Rear Admiral Leslie C. Stevens was selected to be the first chief of the JSPD, although he had limited experience in psychological warfare and covert operations. Stevens, assisted by deputies from each of the other services, initially had a small staff of six officers. The Army concurred in his nomination.[149]

While Paddock was unaware of any previous experience Stevens had in psychological warfare and covert operations, it is hard to envisage the latter being given such an important assignment under the Joint Chiefs of Staff without any prior experience.

Stevens' aeronautical engineering background suggests that he had become aware of, or was directly involved in, the Navy's research into the recovered flying saucer craft from the Los Angeles Air Raid incident, and of the covert intelligence program being run out of Naval Air Station, San Diego. It is more than likely that while Botta was handling the debriefing of the 29 naval operatives out of San Diego, Stevens was involved in other aspects of the same or a similar program run elsewhere by the Navy.

Stevens was awarded a Legion of Merit for:

> ... exceptionally meritorious conduct in the performance of outstanding services to the Government of the United States as Assistant Chief of Staff for Material on the staff of Commander, Air Force, Pacific Fleet, from 3 November 1944 to 5 August 1945. [150]

In such a position, Stevens would have been a key Navy official in assisting the Army's Interplanetary Phenomenon Unit to achieve its objectives. These would include investigating the flying saucer phenomenon in the Pacific theater of the war, and participating in

Figure 42. Document showing Admiral Leslie Stevens involved with Covert Operations

COPY #7
TS 43516

CENTRAL INTELLIGENCE AGENCY
Washington 25, D.C.

Office of the Director

8 January 1951

MEMORANDUM TO: EXECUTIVE SECRETARY
NATIONAL SECURITY COUNCIL

SUBJECT : Draft of NSC Directive on Covert Operations and Clandestine Activities

1. On 14 December 1950, at my request, the National Security Council suspended paragraph 4 of NSC 10/2.

2. I am submitting herewith the draft of a directive for issuance by the National Security Council which clearly defines the responsibilities for covert operations and clandestine activities in peace or in war. This draft was prepared by representatives of this agency in consultation with Rear Admiral Leslie Stevens from the Joint Chiefs of Staff, Brigadier General John Magruder from the Office of the Secretary of Defense, and Mr. Robert Joyce from the Department of State.

3. It is my recommendation that this Directive be sent by the National Security Council to the Departments of State and Defense and the Joint Chiefs of Staff for comment.

4. A related subject which needs clarification is the distinction between covert operations such as may be planned and executed by this agency, and guerrilla warfare conducted by regular forces. I have directed that a paper on this subject be prepared for submission to the NSC.

/s/ WALTER B. SMITH
Director

psychological operations to hide the Unit's covert operations. Consequently, it is almost certain that Admirals Botta and Stevens were early recruits to a covert Navy leadership group that would, beginning in the late 1940's, oversee the design and development of the Navy's space fleet that would in time join an interplanetary alliance. This fleet would be based on similar strategic principles used for the deployment of modern aircraft carrier battle groups.

Being familiar with covert operations and psychological warfare, Admiral Stevens more than likely was instrumental in revealing intimate details of the Navy space fleet to his son, whose own wartime military intelligence service would have made him very familiar with psychological warfare, and even the operations of the Interplanetary Phenomenon Unit. This means that after leaving military service in 1945 to resume his playwriting career, Stevens IV would have in all likelihood continued to have been involved in ongoing psychological warfare operations.

When his father, Admiral Stevens, became the head of "The Joint Subsidiary Plans Division" in 1949, it is realistic to assume that he recruited his son as a military intelligence asset for disseminating key ideas (that were part of psychological warfare operations) into his plays and the film/television industry. According to Knowles, this is a possibility many have long suspected about Stevens:

> Stevens, Sr. died in 1956 but his son was himself an intelligence agent during the Big One and almost certainly maintained his connections after the war. In fact, some believe he remained on the payroll throughout his Hollywood career.[151]

Tompkins confirmed in an interview that Admiral Stevens played a direct role in setting up a secret Navy space program, and that his son, Leslie Stevens IV, was aware of what was happening.[152]

Consequently, we know that prior to his death in 1955, Admiral Stevens revealed significant classified information about the existence of extraterrestrial life and secret space programs to his son. Therefore, Stevens IV was most certainly part of a Navy sanctioned "soft disclosure" initiative, which strategically used the television/movie industry to reveal elements of the Navy's planned secret space program. This would be part of a bigger psychological warfare process that both prepared the public by hiding the truth in plain sight, while also enabling a means of ridicule for those revealing details of the Navy's secret space program without authorization.

Exemplifying this process are the recommendations by the 1953 CIA-sponsored Robertson Panel that issued the Durant Report, which advised the use of the mass media to "educate" the public about flying saucers. The Durant Report recommended debunking flying saucer investigations - by ridiculing the flying saucer phenomenon and the possibility of extraterrestrial life—for national security reasons. The Report stated:

> The "debunking" aim would result in reduction in public interest in "flying saucers" which today evokes a strong psychological reaction. This education could be accomplished by mass media such as television, motion pictures, and popular articles.... Such a program should tend to reduce the current gullibility of the public and consequently their susceptibility to clever hostile propaganda.[153]

Television and movies would be used as part of a psychological warfare program that worked with a number of methods to hide the reality of the flying saucer/UFO phenomenon, and ridicule those investigating it.

A National Security Agency document released by Edward Snowden shows six different ways in which the general public can be deceived through what they are being shown. The document is titled "The Art of Deception: Training for a New Generation of Online Covert Operations," and it contains a graphic illustrating different ways in which the truth can be hidden in plain sight by techniques such as "masking" and "mimicking."[154]

The method of hiding the truth in plain sight would lead to movie and television producers being given partial briefings of classified programs, which would be used for creating new productions. This "soft disclosure" process served the simultaneous goals of creating a means to debunk UFO researchers by critics asserting that their claims were based on popular movies/television series, and also secretly preparing the general public for the truth in case of some catastrophic extraterrestrial disclosure event, e.g., alien motherships appearing over New York City.

Figure 43. NSA Document: Different Ways of Hiding the Truth in Plain Sight

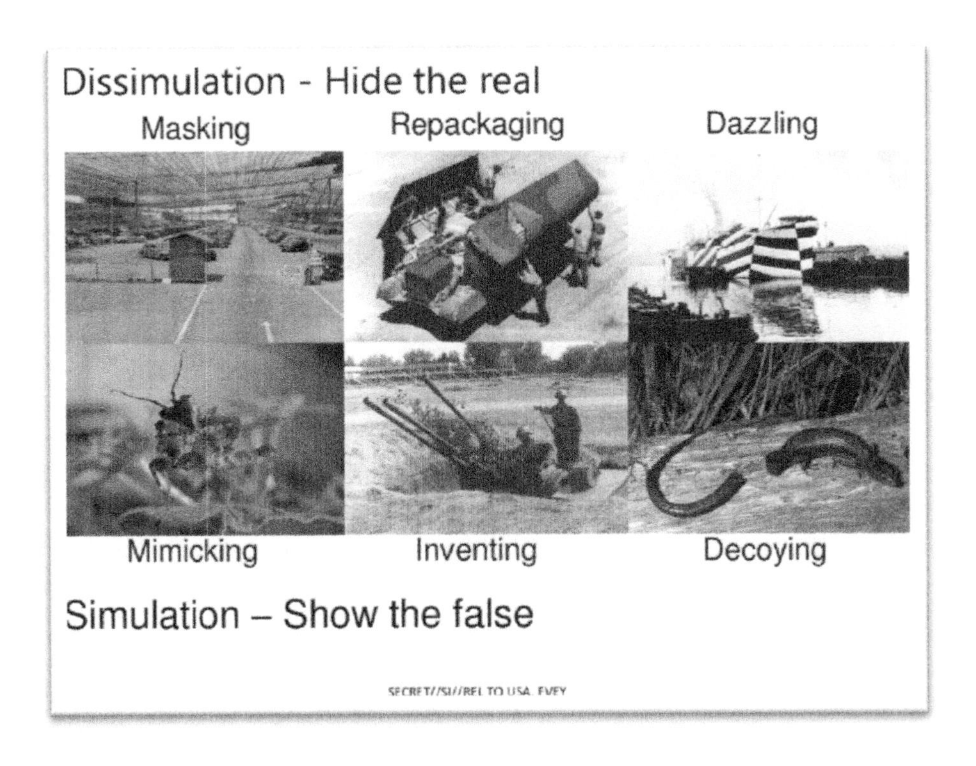

All of this information helps considerably in finally understanding the nature of the business relationship between Roddenberry and Leslie Stevens IV. Stevens could not directly create a science fiction show himself to reveal details of the classified information given to him by his deceased father or from what he had learned while serving with Army Intelligence. Instead, however, it is very plausible that he received permission from Navy officials to share classified information with television/movie producers such as Roddenberry, who then fictionalized the information.

Rodenberry was not alone among television/movie producers with whom Stevens divulged classified information through ideas for new movies or television series. In chapter eight, I examine Stevens relationship with Glenn Larson, creator of yet another famous movie/television franchise—"Battlestar Galactica" —which has a unique place in television history. Notably, on March 17, 2009, its principal actors and script writers were invited to a special forum hosted by the United Nations to discuss lessons learned from the TV series for interspecies conflict resolution.

The business relationship between Gene Roddenberry and Leslie Stevens, suggests that *Star Trek* was initially based on classified information, acquired through Admirals Botta and Stevens, and possibly by the younger Stevens' own direct involvement in military intelligence operations during and after World War II. A future Navy space fleet was to be developed that would actively cooperate with Nordic extraterrestrials in an alliance that countered the Nazis and their Reptilian allies. In the *Star Trek* series, the Nordics were depicted as Vulcans, the Reptilians as Klingons, while the Nazis were represented as genetically enhanced humans.

Ample evidence suggests *Star Trek* was no mere science fiction series, but was actively encouraged through a business arrangement between Roddenberry and the son of a Navy admiral, who had been directly involved in developing a secret US Navy Space Fleet. During the airing of the original *Star Trek* series from

1966-1968, the Navy was taking a significant incremental step towards its secret goal, which at the time was through the Apollo Moon Program.

Endnotes

---

[138] James H. Burns, "Why Roddenberry Created Star Trek," http://file770.com/?p=26855
[139] Chris Knowles, "Secret Star Trek, Part 8: Daystar Trek and the Majestic Nine," http://secretsun.blogspot.com/2013/07/secret-star-trek-part-8-daystar-trek.html
[140] Chris Knowles, "Secret Star Trek, Part 8: Daystar Trek and the Majestic Nine," http://secretsun.blogspot.com/2013/07/secret-star-trek-part-8-daystar-trek.html
[141] Gordon White, "Apolalypse Puja," http://runesoup.com/2013/08/apocalypse-puja/
[142] Gordon White, "Apolalypse Puja," http://runesoup.com/2013/08/apocalypse-puja/
[143] Tom Vallance "Obituary: Leslie Stevens," http://www.independent.co.uk/news/obituaries/obituary-leslie-stevens-1159807.html
[144] Tom Vallance "Obituary: Leslie Stevens," http://www.independent.co.uk/news/obituaries/obituary-leslie-stevens-1159807.html
[145] See Michael Salla, "President Kennedy's deadly confrontation with the CIA & MJ-12 over ET/UFO X-Files," *Exopolitics Journal* 3:2 (July 2009). Available online at: http://exopoliticsjournal.com/vol-3/vol-3-2-Salla.htm
[146] Legion of Merit, http://valor.militarytimes.com/recipient.php?recipientid=307465%5D
[147] Chris Knowles, "Secret Star Trek, Part 8: Daystar Trek and the Majestic Nine," http://secretsun.blogspot.com/2013/07/secret-star-trek-part-8-daystar-trek.html
[148] Private interview, August 24, 2016.
[149] Alfred H. Paddock, Jr., *US Army Special Warfare: Its Origins: Psychological and Unconventional Warfare, 1941-1952* (National Defense University Press, 1982), 78-79. Available online at: http://www.dtic.mil/dtic/tr/fulltext/u2/a118758.pdf
[150] "Military Times Hall of Valor," http://valor.militarytimes.com/recipient.php?recipientid=312320
[151] Chris Knowles, "Secret Star Trek, Part 8: Daystar Trek and the Majestic Nine," http://secretsun.blogspot.co.uk/2013/07/secret-star-trek-part-8-daystar-trek.html
[152] Private interview, August 24, 2016.
[153] Cited from online version of Robertson Panel at: http://www.cufon.org/cufon/robertdod.htm

[154] The Intercept, "The Art of Deception: Training for a New Generation of Online Covert Operations," https://theintercept.com/document/2014/02/24/art-deception-training-new-generation-online-covert-operations/

CHAPTER 6

APOLLO & THE NAVY'S NOVA PROGRAM

On May 25, 1961, President John F. Kennedy gave his famous speech before a joint session of Congress promising to put "a man on the Moon and returning him safely to the earth" by the end of the decade.[155] He called for the development of rockets "much larger than any now being developed, until certain which is superior." Kennedy's speech led to major Congressional funding being provided to NASA for the development of heavy lift rockets to send people and equipment to the Moon. Major aerospace companies immediately began competing for the lucrative contracts to build the heavy lift rockets that NASA would require for the Apollo space program, dedicated to achieving Kennedy's vision.

According to William Tompkins, NASA's Apollo program was only the first stage of a far more ambitious plan that had been developed by the Navy in the early 1950's. While the Army's "Project Horizon" planned to put a dozen soldiers on the Moon for a military/scientific mission by 1966, the Navy planned to have 10,000 people on the Moon by the 1970's as part of a far more extensive manned lunar bases project.[156]

Tompkins says that NASA, and its various civilian space programs such as Apollo, was designed to be a cover for the Navy's secret space program; an arrangement both Presidents Eisenhower and Kennedy supported. The Navy would work through the ostensibly civilian NASA program to achieve its military space objectives, which were considered to be of the utmost national security concern. Consequently, Tompkins has asserted that "NASA is a Naval military organization."[157]

His claim here is supported by NASA Researchers, Richard Hoagland and Mike Bara, the authors of *Dark Mission* who wrote:

> NASA ostensibly is "a civilian agency exercising control over aeronautical and space activities sponsored by the United States"... But contrary to common public and media perception that NASA is an open, strictly civilian scientific institution, is the legal fact that the Space Agency was quietly founded as a direct adjunct to the Department of Defense tasked with specifically assisting the national security of the United States ... It says so right in the original NASA Charter:
>
> > "Sec 305 ... (i) The National Aeronautics and Space Administration shall be considered a defense agency of the United States for the purpose of Chapter 17, Title 35 of the United States Code."[158]

The Navy's manned lunar bases program was part of a four-tiered space program called Nova, which Tompkins states was to be partially implemented through NASA. Nova was designed, from its inception, to enable the Navy to use conventional rocket technologies to establish large bases on the Moon and Mars in the 1960's and 1970's.

According to Tompkins, the Apollo Moon landings were only the first stage of an ambitious four stage plan for Nova. Stage 2 of Nova was to put 10,000 people on the Moon. Stage 3 was to place bases on Mars and other planetary bodies in the Solar System. Finally, Stage 4 was to place manned Navy bases in 12 adjoining star systems.[159]

While Nova would enable the attainment of the first two stages, along with initial elements of the third, it would also require more advanced spacecraft using antigravity and other exotic propulsion systems to achieve all of stage 3, and especially stage 4. These more advanced craft would be secretly designed and developed in the Navy's research facilities at China Lake, California, and then built at a massive underground base in the Wasatch Mountains of Utah, resulting in deep space operations with interstellar travel technologies.

It is an accepted fact that the Nova concept of massive rocket launchers was being studied by two major aerospace companies in the early 1960's that had not been awarded contracts for the production stages of the Saturn heavy launchers for the Apollo Moon missions. According to Encyclopedia Aeronautica:

> General Dynamics (Convair) and Martin Marietta - were given 'consolation' study contracts for Nova in July 1962. Philip Bono of Douglas Aircraft characteristically did his own study without a contract. The contractors were to make preliminary designs of million-pound-payload launch vehicles ... Martin handed in the most comprehensive study, with all possible combinations evaluated... General Dynamics had the most conservative designs, using existing engines or enormous conventional bell-chamber engines in the 3 million pound thrust class. Bono at Douglas characteristically was optimistic about achievable stage mass fractions and had designs with masses considerably less than calculated by the other two contractors.[160]

Citing his involvement with the Douglas study, Tompkins says he and other engineers had utilized various innovative propulsion systems over the years to develop different configurations for the massive Nova rockets and their respective million pound payloads—ten times the payload of the Saturn V rocket:

> We had spent seven years on-and-off with these concepts before NASA put out requests for a bid on the Nova space vehicles (Nova was way before Apollo). Jim Jenkins, Mack Davis, and I spent an enormous amount of time discussing and conceiving these rockets. Our mission in Advanced Design was to analyze the propulsion scheme, develop configurations, trade-off studies, and design massive space freight trucks... I personally designed a 600+ foot high, cone-shaped NOVA vehicle...[161]

The Nova study program officially lasted from 1959 until 1964, when funding was cancelled by NASA due to the successful testing and development of the Saturn V rocket. According to Tompkins, however, Nova continued to be supported behind the scenes by the Navy, which maintained interest as major aerospace contractors submitted unsolicited proposals for Nova's feasibility.

Figure 44. Douglas Design of Nova Rockets

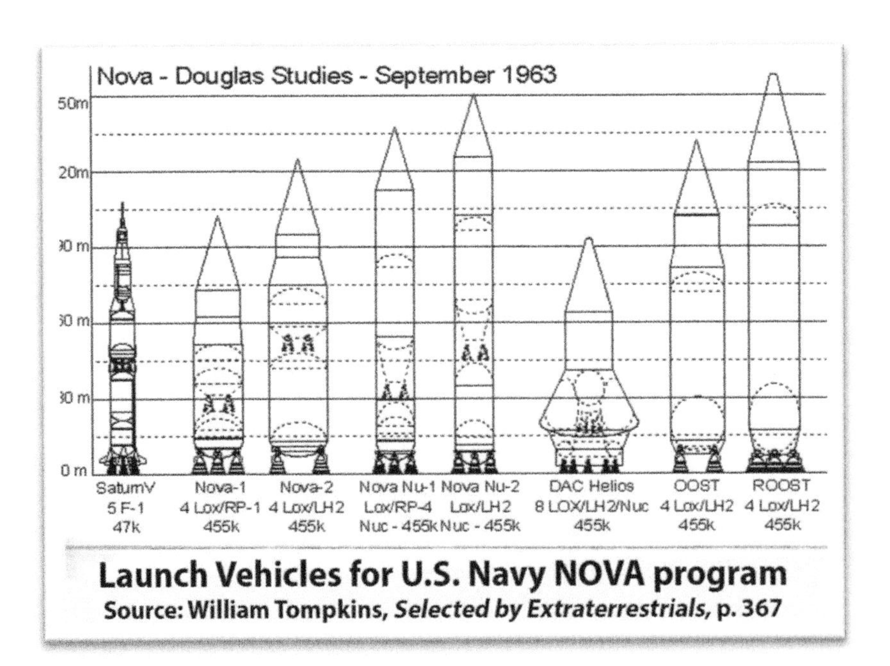

Launch Vehicles for U.S. Navy NOVA program

Source: William Tompkins, *Selected by Extraterrestrials*, p. 367

During this critical early period in the design and development of heavy lift vehicles for the NASA space program, Tompkins' 12 year career at Douglas came to a tumultuous end. It is worth reviewing what happened, and how his dismissal led to him moving over to other major aerospace companies that supported his efforts to get Navy contracts to build the Nova rockets, well after its ostensible cancelation by NASA.

On July 1, 1962, Tompkins took the initiative to go over the heads of the Douglas Aviation Company's manufacturing division to meet with, and establish a personal relationship with the head of NASA's Launch Operations Center (later renamed Kennedy Space Center after the JFK assassination), Dr. Kurt Debus.[162] In 1961, after the Kennedy speech pledging to land a man on the moon by the end of the decade, Debus was chosen to direct the "design, development and construction of NASA's Saturn launch facilities" at Cape Canaveral.[163] He was subsequently appointed as the Space Center's first Director on July 1, 1962; a position he held until his retirement in November 1974.

Tompkins states that he travelled to Cape Canaveral with a detailed model and design proposal for the Saturn launch facilities, together with proposals for building and streamlining the Launch Operations Center. He met with both Dr. Debus and Dr. Werner Von Braun. Debus in particular was so impressed by Tompkins' design and detailed knowledge that he approved his plans on the spot, and had him appointed to a number of NASA committees, including the prestigious Launch Operations Committee Working Group that Debus personally headed. The committee was set up to design and oversee the construction of the Apollo launch complex for both the Saturn and Nova rocket programs. Tompkins recalls what Debus said after their initial meeting:

> I am putting you on my Launch Operations Committee, Complex 39 Facility Planning Committee, and Mission Planning Committee. Then, I would like you to technically consult my Systems Management Committee, Mission

> Control Committee, Facilities and Complex Pane, Launch Operations Working Group, and Checkout Project Office, and Launch Project Office.[164]

Tompkins has provided a Douglas Aviation Company document that corroborates his claim of having been a member of the Launch Operations Committee (L.O.C.) Working Group for the Kennedy Space Center, where he was directly advising Debus on various design proposals. The document refers to Tompkins as "a member of the L.O.C. Facilities Working Group headed by Dr. Debus, Director of L.O.C. for NASA."

Tompkins' personal relationship with Debus led to him securing a $36 million systems engineering contract for the Douglas Aircraft Company.[165] However, by going over the heads of more senior people, he angered leaders of the powerful Manufacturing Division of Douglas. This led to Tompkins being fired, and then the decision was rescinded due to the intervention of his Engineering Division superiors. Finally he was fired again, for a second time, by the Douglas Aviation Company on May 4, 1963.

According to Dr. Robert Wood, who worked at the Douglas Aircraft Company, and its successor McDonnell Douglas, for a total of 43 years, he vividly recalls the firing incident(s) described by Tompkins.[166] He says it was big news at Douglas at the time that someone was fired for getting a large contract with NASA by going over the heads of more senior Douglas officials. Years later, in 2009, Wood would eventually meet Tompkins and decide to help him publish his autobiography by becoming its editor.

In chapter four, it was explained that Nordic extraterrestrials had infiltrated the Douglas Aircraft Company and were covertly assisting Tompkins with the designs of kilometers-long Navy spacecraft. This assistance also included designs for the Nova heavy launch vehicles, and later the Saturn rockets for the Apollo program. His firing incidents were orchestrated, according to Tompkins, by Reptilian extraterrestrials who were controlling key personnel in the Manufacturing Division of Douglas. Their goal was to sabotage Douglas Aircraft Company efforts to assist the Apollo program.

Figure 45. Douglas Memo Showing Tompkins in Debus' NASA Working Group

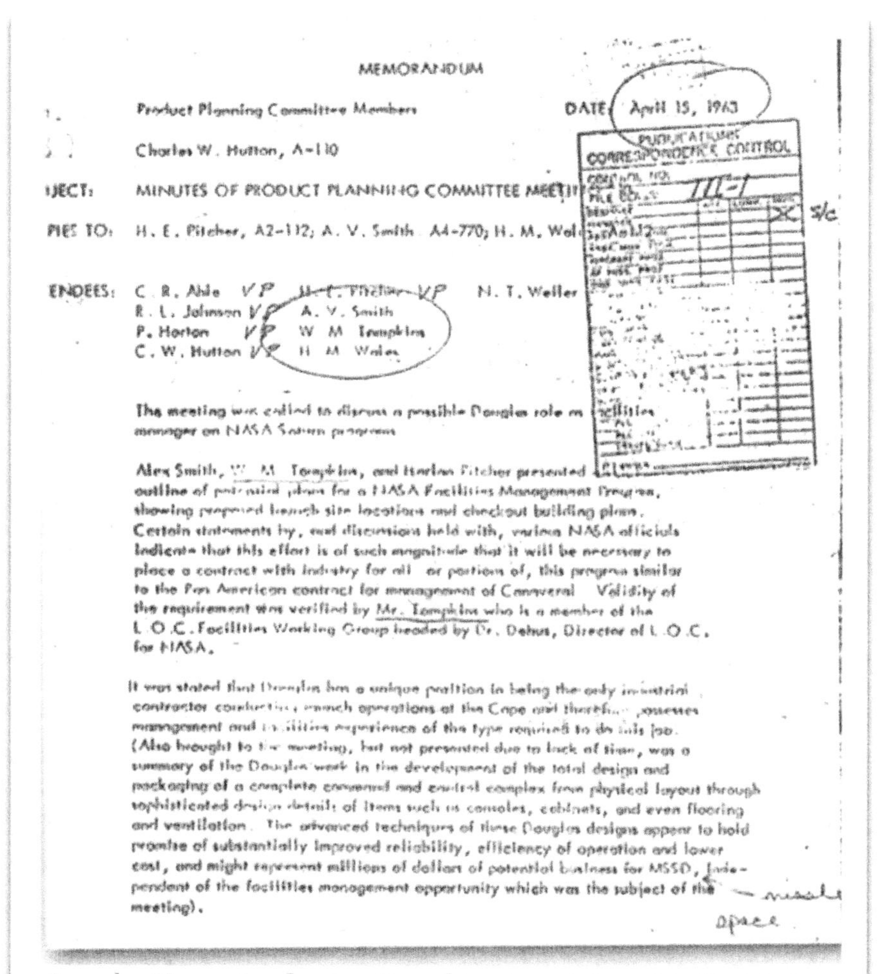

MEMORANDUM

TO: Product Planning Committee Members

DATE: April 15, 1963

FROM: Charles W. Hutton, A-110

SUBJECT: MINUTES OF PRODUCT PLANNING COMMITTEE MEETING

COPIES TO: H. E. Pitcher, A2-112; A. V. Smith A4-770; H. M. Wal[illegible]

ATTENDEES: C. R. Able VP, H. E. Pitcher VP, N. T. Weller, R. L. Johnson VP, A. V. Smith, P. Horton VP, W. M. Tompkins, C. W. Hutton VP, H. M. Wales

The meeting was called to discuss a possible Douglas role as facilities manager on NASA Saturn program.

Alex Smith, W. M. Tompkins, and Harlan Pitcher presented an outline of potential plans for a NASA Facilities Management Program, showing proposed launch site locations and checkout building plans. Certain statements by, and discussions held with, various NASA officials indicate that this effort is of such magnitude that it will be necessary to place a contract with industry for all or portions of, this program similar to the Pan American contract for management of Canaveral. Validity of the requirement was verified by Mr. Tompkins who is a member of the L.O.C. Facilities Working Group headed by Dr. Debus, Director of L.O.C. for NASA.

It was stated that Douglas has a unique position in being the only industrial contractor conducting launch operations at the Cape and therefore possesses management and facilities experience of the type required to do this job. (Also brought to the meeting, but not presented due to lack of time, was a summary of the Douglas work in the development of the total design and packaging of a complete command and control complex from physical layout through sophisticated design details of items such as consoles, cabinets, and even flooring and ventilation. The advanced techniques of these Douglas designs appear to hold promise of substantially improved reliability, efficiency of operation and lower cost, and might represent millions of dollars of potential business for MSSD, independent of the facilities management opportunity which was the subject of the meeting).

PUBLICATIONS CORRESPONDENCE CONTROL

III-1

Douglas Memo Confirming Tompkins Appointment to Kurt Debus Working Group for Designing Launch Operations Center

Source: William Tompkins, *Selected by Extraterrestrials*, p. 351

After noting the disappearance of two engineers from the Engineering Division, Tompkins says he told a colleague:

> There must be Reptilians at the very top of manufacturing management in Douglas. They are preventing us from implementing all of our engineering proposals that I have presented to Douglas Corporate to manage all of the NASA Moon and Mars mission programs.[167]

Offering a further explanation, Tompkins reveals that the Reptilians could infiltrate Douglas by using advanced technologies that made them appear physically human. He said: "some of the Nordic aliens in the Douglas shop whom I thought were helping us, may be disguised Reptilians, and this could be happening now in NAA [North American Aviation], too."[168] Subsequently, Tompkins summarizes the complex situation involving different extraterrestrial groups that had infiltrated different companies to either help or sabotage the Navy and Apollo programs:

> My main concern ... is why the hell are the Nordics pushing us to the Moon, but those damn Reptilians, who seem to look like the Nordics, are throwing wrenches into the gears?[169]

Furthermore, Tompkins contends that Debus was very knowledgeable about the Nordic extraterrestrials that had been helping Maria Orsic establish the first German Vril flying saucer prototypes.[170] Debus was himself, according to Tompkins, telepathically in touch with Nordic extraterrestrials who were working behind the scenes to help the Apollo Program succeed. In a later interview, Tompkins described how his initial meeting with Debus was orchestrated by his Nordic extraterrestrial secretary from Douglas, who made it possible for him to overcome many obstacles in order to meet with Debus.[171]

> My secretary had said before we even went down there that I've written you a letter which will introduce you to the top people in NASA, but you're never going to use it. So then she did all of that stuff where I'm driving into that zone arsenal, the secret area, and the secret doors, gates, opened without stopping me. The armed guards don't grab me and shoot me... And all of these things happened exactly as she told me it would.
>
> She also told me not to believe a lot of the stuff that the German guys would say. But Dr. Debus, to me, even though von Braun was the top guy, Debus somehow immediately, right after I started talking, became a friend. And I can't describe it any other way. And virtually, I couldn't say anything wrong.[172]

The connection that both Tompkins and Debus had with Nordic extraterrestrials was a key factor in why they established a strong personal rapport from their first meeting.

After being sacked for the final time by Douglas in May 1963, it was Debus who put Tompkins in touch with his next employer, North American Aviation (NAA). Tompkins summarizes the situation as follows:

> Because of my unsolicited, major change to the entire Apollo Moon proposal a year earlier, I was fired at Douglas. This was for going over corporate heads. My [July 1, 1962] presentation was to the two most important space people on the planet. Dr. Kurt Debus and Dr. von Braun, the head of NASA... Dr. Debus was so impressed with my depth of [knowledge of] the alien threats and their implications with the Apollo and Naval missions that he appointed me to his Mission Launch Control and Facilities Planning Committees. A call to Dr. Debus and I was on board the next week at North American Aviation, Rocketdyne, a NAA division and

> Nucleonics, also a division of NAA, at a director level in advanced technologies research. [173]

From 1963 to 1966, Tompkins worked at the Rocketdyne Division of North American (now part of Boeing), and took with him many of the advanced spacecraft designs he had developed at Douglas.

During his Rocketdyne employment, Tompkins relates how he successfully persuaded management to replicate what he had been doing at Douglas in terms of sending unsolicited proposals to the Navy to win contracts for classified programs.[174] Tompkins got Rocketdyne managers to approve an unsolicited bid to build Nova heavy lift vehicles based on the old Douglas Aircraft proposals he had previously worked on.

Tompkins says that he was now being assisted by two different female Nordic extraterrestrials, other than those at Douglas. They were helping him solve the complex engineering design problems for building large rocket systems such as Nova and the Saturn V, which he also worked on at Rocketdyne. He wrote that the two Nordics were "aliens assigned here to see that the Apollo Moon program will accomplish its goals."[175]

Tompkins prepared a business plan for Rocketdyne which incorporated some of the innovative propulsion technologies he had studied at Douglas for interstellar travel, as well as more conventional rocket technologies for Nova missions throughout the solar system:

> These included interplanetary, manned Mars, Venus, Mercury and deep space, major solar system planets, or their habitable moons. I conceived and they developed naval bases on all of them... the availability of the improved NOVA truck vehicle rockets into naval scout ships will considerably increase the effectiveness of early solar system, manned naval planetary Moon, Mars fly-by and station missions. It will also increase development of the naval Mars base operation. [176]

This culminated in a detailed systems engineering plan for colonizing our own and adjacent solar systems. The North American systems plan was to use the powerful Nova rockets as a critical aspect for colonization. Tompkins says that his lack of enthusiasm for incorporating nuclear propulsion systems into his designs for future Navy spacecraft came to be resented by senior figures at Rocketdyne.

During this time, one of his old Douglas bosses encouraged him to move over to TRW where his advanced design and engineering skills would be better utilized:

> Noel Crates, my old Douglas Delta-II Thor DM-18 missile Boss who had resigned Douglas after I was fired, had accepted a research position at the TRW space Think Tank at Redondo Beach, near LAX.... For years, Noel constantly tried to get me to switch over from North American Rocketdyne to TRW saying, "Bill, this is the center of the Galaxy where the action is, it is all happening at TRW. You must get on the staff here. There is a lot of that they are not aware of, for some reason, you do understand. There are literally hundreds of classified projects that need your big picture evaluation. Bill, in many ways you are light years ahead of them.[177]

Come 1966, Tompkins' interest in antigravity over nuclear propulsion led to him being booted out of Rocketdyne's Nucleonics, and so he finally took up his friend's offer to move over to the greener pastures of TRW (acquired by Northrup Grumann in 2002). Tompkins spent the next four years (1967-1971) at TRW, and says he found much support for his innovative design and engineering ideas for different space propulsion systems at TRW's futuristic campus at Redondo Beach:

> TRW is the most fantastic space Think Tank campus at the tip of this southwest arm of the galaxy. It has been

> addressing extraterrestrial threats, conceiving Naval space battle groups and weapons to process these threats since 1960.[178]

Due to his background in Naval intelligence and success in getting major Navy contracts when working at Douglas and Rocketdyne, Tompkins was given multiple TRW projects to work on.[179] This included a Nova proposal by TRW, where the company eventually won the much sought after Navy contract. Tompkins wrote:

> After extensive evaluation of hundreds of research projects, involving every possible concept to unconventional space population and 14 major classes of Naval space ships, from fighter to 2.5 kilometer-long intergalactic battle cruisers, an unsolicited system development plan was conceived and submitted to ONI [Office of Naval Intelligence] and ONR [Office of Naval Research], along with countless Apollo/NOVA lunar, planet and star missions projects. I accomplished those missions, several years later, at TRW. [180]

However, Tompkins said that the covert Navy plan for developing Nova in order to place 10,000 personnel on the Moon came to a crushing end during the Apollo 11 mission. At the time of the Apollo Moon landing, Tompkins was working for TRW, which was responsible for a range of critical components used in Apollo spacecraft. TRW personnel received the same telemetry viewed at the Launch Operations Center/Kennedy Space Center for Apollo Moon missions. As a TRW employee, there is no reason to doubt Tompkins' account of being physically present among a large contingent of staff at TRW's Redondo Beach "Space Park" headquarters to witness the Moon landing.

During the Apollo 11 Moon landing in July 1969, Tompkins states that television cameras from the Apollo Lander provided a live feed of what was being witnessed by Armstrong and Aldrin. Next, Tompkins explains how Armstrong and Aldrin were met by a fleet of extraterrestrial spaceships that were ominously close to the Apollo Lunar Lander:

> The Landing Module (LEM) actually impacted the Moon surface in the Sea of Tranquility Crater, which had tremendous size vehicles parked around part of its rim. When astronaut Neil Armstrong made that First Step for Man on the Moon he looked up to the edge of the crater and said to mission control: "There are other ships here, they are enormous. The public did not hear that statement or see the massive alien starships. Armstrong panned his camera in a 360 degree motional all around the crater and the CIA then classified the information as way above top secret.[181]

Tompkins drew a picture of what he, along with other TRW and NASA personnel, witnessed on the Apollo live feed camera (see Figure 46).

Incredibly, Tompkins says an alien voice was clearly heard over the Apollo 11 radio transmission: "Finish a total of six of your Apollo missions; take your photos, pick up some rocks, go home, and don't come back."[182] How this threatening action by one or more extraterrestrial groups effectively put an end to the Apollo program, as well as the Navy's covert Nova program for manned Moon bases, is described by Tompkins this way: "The NO TRESPASSING sign went up. With the program over, everybody went home, and nearly all 400,000 NASA and contractor employees got the pink slip."[183]

What Tompkins clearly outlines he saw happen via the NASA live feed of the Moon landing is consistent with HAM radio intercepts of the live feed, and the claims of former NASA employee Otto Bender. According to Bender, the Apollo 11 astronauts transmitted that they were being watched by large alien vehicles.

Figure 46. UFOs Tompkins Witnessed on NASA Live Feed During Apollo 11 Landing

Wiliam Tompkins Drawing of NASA Live Feed of Starships on Edge of Sea of Tranquility Crater Menacing Apollo 11 Astronauts
Source: William Tompkins, *Selected by Extraterrestrials*, p. 417

Bender confirmed that HAM radio operators had indeed intercepted VHF signals transmitted from Apollo 11 to NASA's Houston headquarters with the following message, which NASA screened from the public:

> Mission Control: What's there ? Mission Control calling Apollo 11.
>
> Apollo 11: These babies are huge, sir ... enormous....Oh, God, you wouldn't believe it! I'm telling you there are other space craft out there... lined up on the far side of the crater edge... they're on the moon watching us.[184]

On August 27, 2012, Dr. Steven Greer, Founder of the *Disclosure Project*, revealed that he spoke with close relatives of both

Armstrong and Aldrin who had been told the truth about what the astronauts saw on the moon:

> Close friends and very close family members of both Neil Armstrong and Buzz Aldrin have separately told me that indeed there were numerous, large UFOs around the crater where the Lunar Module landed and that these were seen by both Armstrong and Aldrin. I have also spoken to military officers that have seen the footage of this event – but it has never been made public.[185]

Greer's testimony is significant since it supports Tompkins' claim that there was a live video feed recording what Armstrong and Aldrin were seeing.

So why did NASA eventually terminate the Apollo missions if extraterrestrial visitors were there and watching the Earth? The answer according to Armstrong, as relayed by an unnamed Professor at a NASA symposium, is as follows:

> Professor: What really happened out there with Apollo 11?
>
> Armstrong: It was incredible ... of course, we had always known there was a possibility ... the fact is, we were warned off. There was never any questions then of a space station or a moon city.
>
> Professor: How do you mean "warned off"?
>
> Armstrong: I can't go into details, except to say that their ships were far superior to ours both in size and technology – Boy, where they big! ... and menacing .... No, there is no question of a space station.
>
> Professor: But NASA had other missions after Apollo 11?

> Armstrong: Naturally – NASA was committed at that time, and couldn't risk a panic on earth.... But it really was a quick scoop and back again.[186]

The unnamed Professor was first quoted by Timothy Good in his ground breaking book, *Above Top Secret,* and he corroborates Tompkins' claim that NASA was verbally warned off the moon, and that there were plans to build a "moon city." The city was in fact planned to be a Navy base that was part of the Nova program.

The extraterrestrial occupants of the large starships intimidating the Apollo 11 mission did not want the US Navy establishing a beachhead for future military bases on the Moon. By preventing the Navy from moving forward with its plan to put 10,000 people on the Moon using a number of Nova rocket launchers throughout the 1970's, the Nova program effectively came to a crushing end in July 1969.

It would take the Navy more than a decade before it could complete the construction of its first antigravity space vehicles as part of its secret space program. This was achieved with the covert assistance of Nordic extraterrestrials, despite the sabotaging efforts of Reptilians embedded within the military industrial complex. According to Tompkins and other whistleblowers, the first Navy space battle groups were deployed in the early 1980's during the Reagan Administration, thereby establishing a US Navy presence in deep space for the first time.[187] During this period, Tompkins retired from the aerospace industry. In the late 1980's, he began a new mission of covertly preparing Navy personnel and their children for a future where open contact with extraterrestrial life, and knowledge of the secret space programs, would become widespread.

Endnotes

---

[155] President John F. Kennedy, "Excerpt from the 'Special Message to the Congress on Urgent National Needs'," May 25, 1961. https://www.nasa.gov/vision/space/features/jfk_speech_text.html#.WA-MyaLvRBw
[156] United States Army, *Project Horizon: Volume II: Technical Considerations & Plans* http://www.history.army.mil/faq/horizon/Horizon_V2.pdf
[157] William Tompkins, *Selected by Extraterrestrials* (Createspace, 2015) p. 408.
[158] Richard Hoagland and Mike Bara, *Dark Mission: The Secret History of NASA* (Feral House Book, 2007), p. II.
[159] William Tompkins discussed NOVA and its goals in *Selected by Extraterrestrials*, 366-67.
[160] "Nova" Encyclopedia Astronautica, http://www.astronautix.com/n/Nova.html
[161] William Tompkins, *Selected by Extraterrestrials* (Createspace, 2015) 31-32.
[162] In Tompkins autobiography, Dr. Robert Wood, the editor, says that he has a NASA memorandum dated August 28 (1962) that such a meeting occurred. *Selected by Extraterrestrials,* 339.
[163] Wikipedia, https://en.wikipedia.org/wiki/Kurt_H._Debus
[164] William Tompkins, *Selected by Extraterrestrials*, p. 338.
[165] William Tompkins, *Selected by Extraterrestrials*, p. 344.
[166] Dr. Robert Wood revealed this in a personal conversation on October 11, 2016 during a meeting with William Tompkins in San Diego. Biographical information on Dr. Wood is available at: http://majesticdocuments.com/team/robertwood.php
[167] William Tompkins, *Selected by Extraterrestrials*, p. 353.
[168] William Tompkins, *Selected by Extraterrestrials*, p. 370.
[169] William Tompkins, *Selected by Extraterrestrials*, p. 371.
[170] William Tompkins Interview with Jeff Rense, July 8, 2016, http://rense2.gsradio.net/rense/special/rense_070816_hr2.mp3
[171] See William Tompkins Interview with Jeff Rense, July 8, 2016, http://rense2.gsradio.net/rense/special/rense_070816_hr2.mp3
[172] "Rense Radio Interview with William Tompkins and Maj. George Filer & Frank Chille," March 23, 2016, http://spherebeingalliance.com/blog/our-technology-decades-ahead-of-whats-known-part2.html
[173] William Tompkins, *Selected by Extraterrestrials*, pp. 355-56.
[174] William Tompkins, *Selected by Extraterrestrials*, pp. 368-69.
[175] William Tompkins, *Selected by Extraterrestrials*, p. 363.

---

[176] William Tompkins, *Selected by Extraterrestrials*, p. 397.
[177] William Tompkins, *Selected by Extraterrestrials*, p. 399.
[178] William Tompkins, *Selected by Extraterrestrials*, p. 405.
[179] In his autobiography, Tompkins describes 72 projects he was working on while at TRW, William Tompkins, *Selected by Extraterrestrials*, p. 407.
[180] William Tompkins, *Selected by Extraterrestrials*, p. 398.
[181] William Tompkins, *Selected by Extraterrestrials*, p. 418.
[182] William Tompkins, *Selected by Extraterrestrials*, p. 415.
[183] William Tompkins, *Selected by Extraterrestrials*, p. 415.
[184] "Bad Moon Rising," http://www.thelivingmoon.com/46roslin_gate/01archives/BadMoonRising.htm
[185] Dr. Steven Greer, "Neil Armstrong's UFO Secret,"http://tinyurl.com/h7csgpz
[186] Timothy Good, *Above Top Secret: The Worldwide U.F.O Cover-Up* (Quill, 1989) p. 186
[187] See Michael Salla, *Insiders Reveal Secret Space Programs and Extraterrestrial Alliances* (Exopolitics Institute, 2015).

CHAPTER 7

EXTRATERRESTRIAL "SPECIAL PROJECTS" AT THE NAVY LEAGUE

William Tompkins has revealed that from the late 1980's to 1999, he was given permission to create a number of "Special Projects" involving extraterrestrial life and advanced technologies at a council of the Navy League of the United States. The purpose was to use this civilian organization as a vehicle for educating retired Navy officers and their children about the reality of these pressing issues.

The Navy League is a national organization with approximately 40,000 members in more than 220 councils around the world. It provides support for the Navy and its sister sea services. Founded in 1902, it launched with the encouragement of President Theodore Roosevelt who said: "Give hearty support to the policy which the Navy League is Founded to Further."[188] The Navy League describes itself as: "the foremost citizens' organization to serve, support and stand with all the sea services—the US Navy, US Marine Corps, US Coast Guard and US-flag Merchant Marine."[189]

Tompkins created the Navy League Rogue Valley Council in Medford, Oregon in 1991, and became its inaugural President. His prior work for Naval Intelligence and leading aerospace contractors, from 1942 to 1984, had led him to the conclusion that the Navy League had an important role to play when it came to the topic of extraterrestrial life and advanced technologies. This was a conclusion that was shared by senior officials in the Navy League and his contacts in Naval Intelligence.

Previous chapters have discussed Tompkins participation in various aspects of the Navy's secret space program as it went from the early design phase in the 1950's, leading to detailed architectural and systems engineering plans for their construction in the 1970's, and finally to the space battle groups being deployed in the early 1980's. Coinciding with the deployment of the Navy's secret space battle groups, Tompkins "officially retired" from the aerospace industry in 1984. He then began a new phase in his already remarkable career. Tompkins would now educate select Navy officers and their children about extraterrestrial life and technology to prepare them for a world where the existence of alien life and an interstellar Navy space program became common knowledge.

In January 1984, Tompkins moved to Medford, Oregon where a high concentration of reserve and retired Navy officers resided. He says that he later attended a Navy League National Convention in Seattle, Washington. Conversations with senior Navy officers and members of the Navy League led to him receiving support for a plan to develop a series of Navy led "Special Projects" dealing with the extraterrestrial topic.

Tompkins wrote a letter to Rear Admiral Larry Marsh, Commander of Naval Base Seattle and head of Submarine Group Nine at Bangor, Washington (from September 1990 to June 1992), requesting permission to have a group of retired Navy personnel take a ride on a submarine. Tompkins says he also let the Admiral know of his plans to form a Navy League Council in Medford.

The Navy is very careful when it comes to who it allows on its nuclear submarines, according to Tompkins, and does strict security checks. Presumably, after doing a security check on Tompkins, Admiral Marsh decided to not only grant permission for the submarine ride, but he would also take what Tompkins describes as a *rare step,* by personally traveling down to Medford and presiding over the inauguration ceremony for Medford's Rogue Valley Navy League Council.

Public records verify that the Admiral did travel to Medford where he met with Tompkins and the other retired or reserve Navy officers, and formally participated in the launch of the Rogue Valley Council Navy League. A photo appears below of Tompkins meeting with Admiral Marsh at the opening ceremony for the Council in 1991.

Figure 47. Bill Tompkins & Admiral Larry Marsh at Opening of Navy League Rogue Valley Council in 1991

Later, Tompkins and fellow Navy League officials travelled to Seattle for their submarine ride, and he reports that Admiral Marsh privately met with him there too. Tompkins recalls specifically discussing the extraterrestrial phenomenon and UFO sightings in the Seattle area with the Admiral. Marsh, according to Tompkins, in turn related some of his own direct knowledge of extraterrestrial life and technology, and how this was being developed for a Navy secret space program. Tompkins further states that Admiral Marsh had been briefed about the Navy's secret space program, and approved the "Special Projects" that Tompkins planned to implement at the Medford Navy League Council.

It's worth pointing out that service on submarines has many similarities to service in space. Indeed here, Tompkins points out that the first Navy prototype spacecraft were in fact converted submarines. In an interview, he explained that these prototypes were first developed in the 1950's using submarines:

> And at Douglas, in the secret think tank, we were looking at every type of space vehicle we would need to go out into the galaxy, then our U.S. Navy. So submarines came up and we discussed this. We said, "That's the easiest thing, the quickest way, we can get out there. We'll just take a regular Navy submarine, pull out the whole nuclear system, put in the anti-gravitational system, and we're going to use these right away."[190]

Historical records show that the first nuclear powered submarine, the USS *Nautilus* (SSN-571), became operational in 1955.[191] This makes Tompkins' scenario possible, since it was feasible for a "Nautilus class" of spacecraft to be based on converted nuclear powered submarines. Tompkins says that converted submarines

Figure 48. First "Nautilus Class" Nuclear Submarine built for Navy by General Dynamics

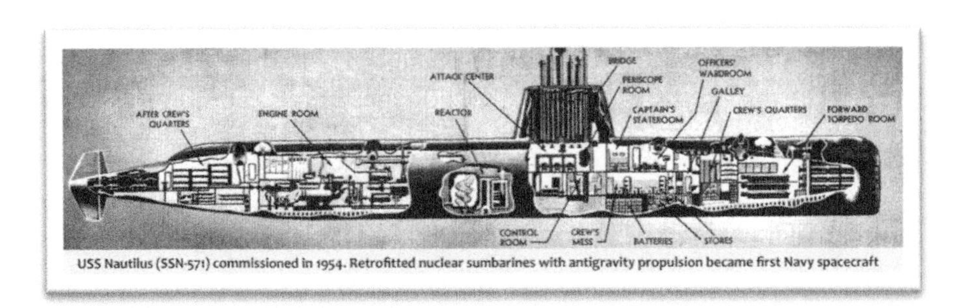

USS Nautilus (SSN-571) commissioned in 1954. Retrofitted nuclear sumbarines with antigravity propulsion became first Navy spacecraft

were first deployed in space in the 1970's as a prelude to the deployment of the larger kilometers-long spacecraft.[192]

As the commander of the Submarine Group Nine facility, it is probable that Admiral Marsh would have at least known of the idea of "Nautilus class" spacecraft, especially considering that training of personnel would have been very similar to the conventional training of crews on nuclear powered submarines. This connection would support Tompkins' claim that Marsh was familiar with the Navy's secret space program, and could have revealed some of this in their second meeting.

Figure 49. Tompkins with Navy League members who took a ride on the Nuclear Powered Submarine USS Alaska SSBN 732

Of critical importance, if Tompkins was helping to prepare Sea Cadets associated with the Navy League for future space operations with the Navy's secret space program, "Solar Warden," then their training would have been invaluably assisted by instructors who were familiar with the conventional training of crews for modern nuclear powered submarines.[193] Tompkins emphasized in an interview that the submarine visit was not merely a joy ride, but was a vital component of one of the "Special Projects" to be undertaken at the Rogue Valley Council in training personnel for future space missions.[194]

Tompkins later met with Rear Admiral Hugh Webster, President of the San Diego Council and corporate director at the national level of the Navy League. During their meeting, Tompkins says that he shared with Webster some of his knowledge about extraterrestrial projects he had been involved in during his Navy and aerospace career. Admiral Webster, prior to his 1991 retirement, was Commander of the Operational Test and Evaluation Force.[195] Previously he had served as Commander of the Naval Surface Group Western Pacific (Commander Task Force (CTF) 73/CTF 75).[196] On November 1, 1986, Admiral Webster led the first visit to China in 40 years. From 1978 to 1980, he was commander of the USS England DLG/CG-22.

Moreover, Tompkins says he also received Admiral Webster's full support for his plans to develop a series of Navy led "Special Projects" dealing with the alien topic. Accepting Tompkins' testimony, this means that two navy admirals, Marsh and Webster, were familiar with the "Special Projects" he was leading at Medford, Oregon.

Later, in 2001, Tompkins says Admiral Webster gave him permission to disclose all he knew about the Navy's involvement in extraterrestrial related projects in his planned autobiography. On the back cover of *Selected by Extraterrestrials,* Tompkins recalls the conversation he had with Webster:

Early in 2001 I called on Admiral Hugh Webster, Navy League Corporate Director, Washington DC and San Diego CA. We had a five hour meeting on my ongoing book–writing concerning the extraterrestrial threats to our planet. After Admiral Webster read portions of my document and backup technical documentation, I asked Hugh, "How much of this can I include in a published book?" He said. "Bill; TELL IT ALL. This is most important to our country. Don't leave anything out."[197]

Figure 50. Rear Admiral Hugh L. Webster & Admiral Xinchun Ma celebrating the first visit by US Navy ships to China in 40 Years

As the inaugural President of the Rogue Valley Council, Tompkins was soon after promoted to Vice President of the Navy League for the entire State of Oregon. He has supplied documents confirming that he was indeed assigned to these very senior positions within the Navy League, despite not having previously been a member. This is highly unusual and would only have occurred through the approval of top senior figures at the national leadership for the Navy League, such as Admirals Marsh and Webster, as Tompkins claims.

In his autobiography, Tompkins describes his main mission in creating the Navy League Council in Medford:

> Created Southwest Navy League Council [Medford] with 37 top Navy officers, the "special projects" group, pilots from various employers, all working on alien interaction projects."[198]

Later, in an interview in March 2016, Tompkins further added:

> You've got a hand out of Navy people responsible for this Navy League Council that I started up in Oregon, and we had specific missions up there with extraterrestrials, particularly in large mountains up there. It's interesting that if you read that little list of names how many of these personnel, who were out of the Navy, but are still in the Reserve, are pilots, Navy pilots, Commanders or Captains. I happened to have one of the largest groups in the country in this Navy League, supporting extraterrestrial missions.[199]

One of his immediate tasks was to set up a Command, Communications and Control system for the Navy League's Medford Council. To achieve this, Tompkins set up the "Sea Cadets" in which junior high school children of navy officers and others would be given a basic education of naval maritime and aerospace operations.

Tompkins was able to secure 14 computers from the Miramar Naval Air Station (previously used for F-14 flight training) for this purpose. He says that the command and control system (CIC), which was set up for the Sea Cadets, was planned for use as a means of educating Navy officers and "sea cadets" about extraterrestrial and secret space operations.

Figure 51. Letter thanking Miramar NAS for computer systems for CIC Simulator

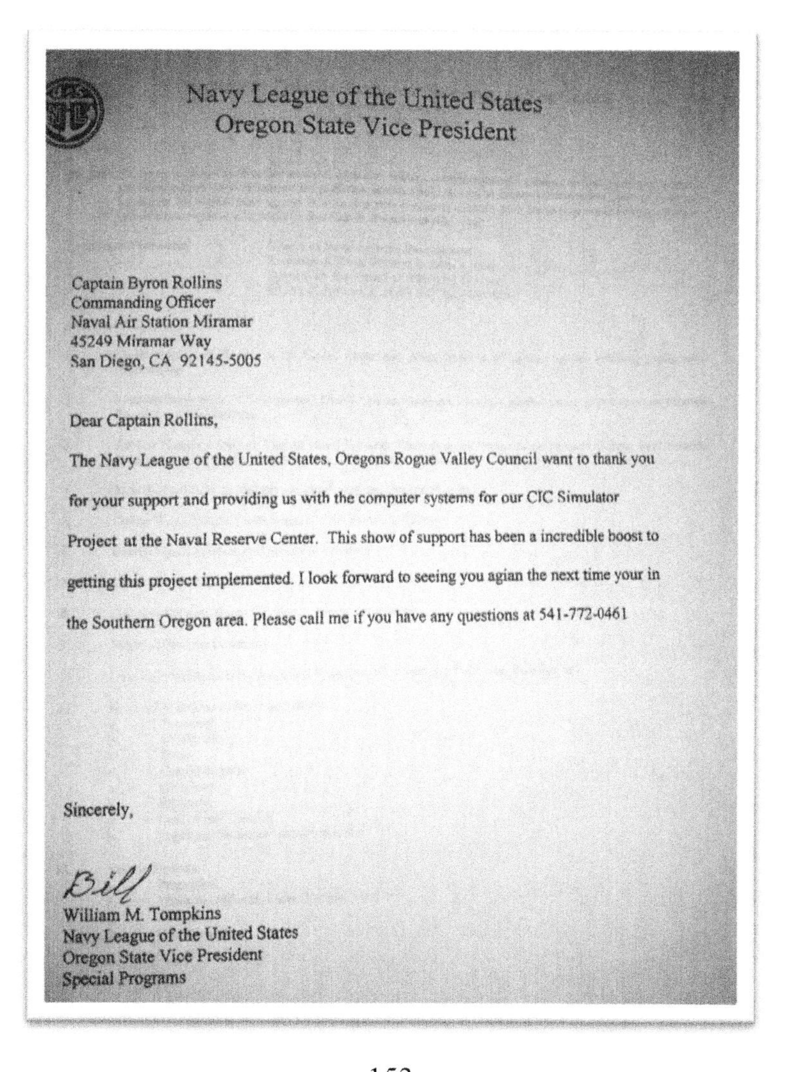

Navy League of the United States
Oregon State Vice President

Captain Byron Rollins
Commanding Officer
Naval Air Station Miramar
45249 Miramar Way
San Diego, CA 92145-5005

Dear Captain Rollins,

The Navy League of the United States, Oregons Rogue Valley Council want to thank you for your support and providing us with the computer systems for our CIC Simulator Project at the Naval Reserve Center. This show of support has been a incredible boost to getting this project implemented. I look forward to seeing you agian the next time your in the Southern Oregon area. Please call me if you have any questions at 541-772-0461

Sincerely,

Bill

William M. Tompkins
Navy League of the United States
Oregon State Vice President
Special Programs

Significantly, Tompkins has supplied documents to support his remarkable claims. One of these documents confirms the computer systems given to him by the Miramar Naval Air Station (see Figure 51).

Another document is a hand out prepared by Tompkins for the Navy League, describing the deep space and extraterrestrial operations project being set up at the Navy League. He refers to the operations system that was being set up as a "Special Project."

Figure 52. Special Projects for Inter-Galactic Operations

Yet another document confirms that "Special Projects" were indeed being discussed and conducted at the Navy League in Medford, as Tompkins claims. This document is the minutes of a meeting held on October 5, 1993. It confirms that Tompkins was President of the Rogue Valley Council [Medford].

Most importantly, the "Minutes" confirm that an item on the agenda was a report by a "Special Projects Committee." Tompkins says that this Committee was in charge of six "Special

Projects" with each having their own sub-committee. One was the "Inter-Galactic Operations Special Project" according to the previous document.

Figure 53. Minutes of Meeting Featuring Special Projects Committee

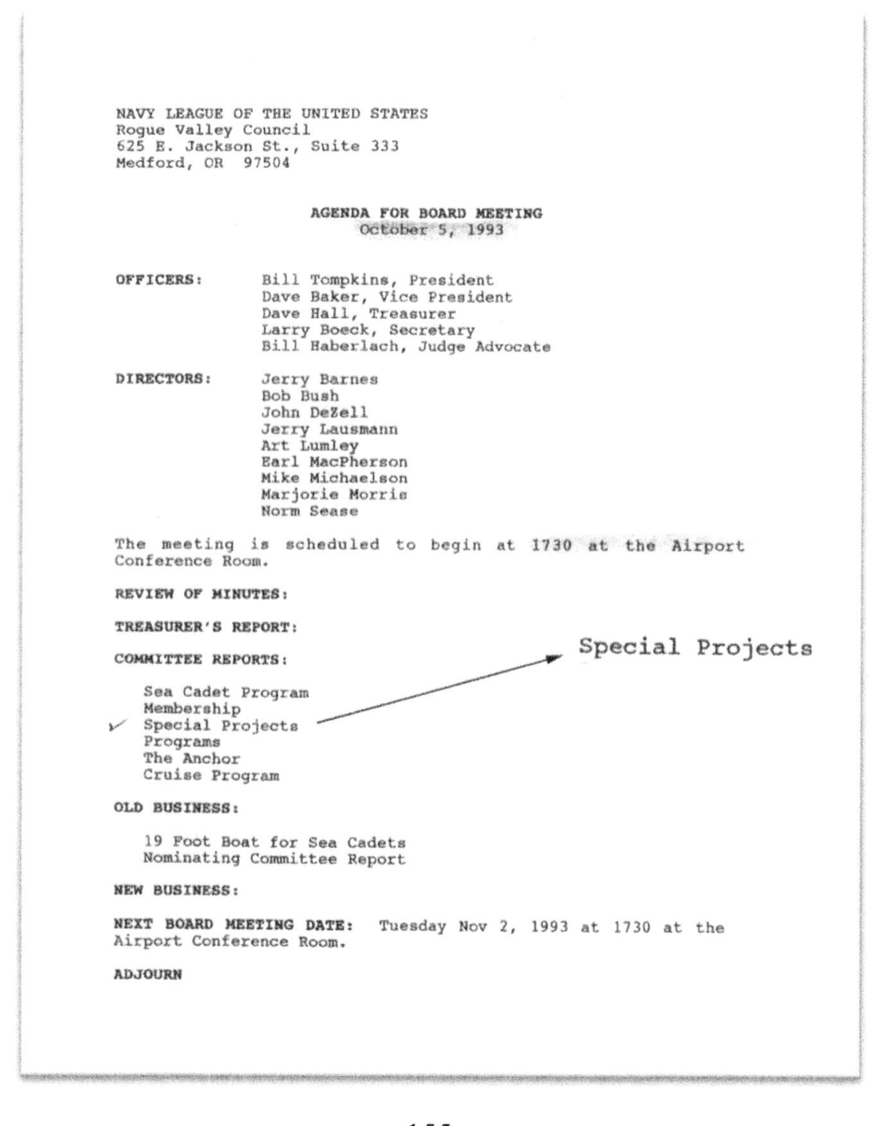

NAVY LEAGUE OF THE UNITED STATES
Rogue Valley Council
625 E. Jackson St., Suite 333
Medford, OR 97504

AGENDA FOR BOARD MEETING
October 5, 1993

OFFICERS: Bill Tompkins, President
Dave Baker, Vice President
Dave Hall, Treasurer
Larry Boeck, Secretary
Bill Haberlach, Judge Advocate

DIRECTORS: Jerry Barnes
Bob Bush
John DeZell
Jerry Lausmann
Art Lumley
Earl MacPherson
Mike Michaelson
Marjorie Morris
Norm Sease

The meeting is scheduled to begin at 1730 at the Airport Conference Room.

REVIEW OF MINUTES:

TREASURER'S REPORT:

COMMITTEE REPORTS:

Sea Cadet Program
Membership
✓ Special Projects
Programs
The Anchor
Cruise Program

OLD BUSINESS:

19 Foot Boat for Sea Cadets
Nominating Committee Report

NEW BUSINESS:

NEXT BOARD MEETING DATE: Tuesday Nov 2, 1993 at 1730 at the Airport Conference Room.

ADJOURN

In private interviews, Tompkins has described some of the other "Special Projects" that he says he was directly involved in. One involved reports of UFO's seen entering and leaving Mount Shasta, which was believed to be an extraterrestrial base. Pilots affiliated with the Navy League would fly around the mountain to find entrances, and all-terrain vehicles were employed for the same purpose.

Another "Special Project" dealt specifically with programs on "extended life." Tompkins says that while he worked at TRW from 1967 to 1971, one of the programs he was assigned to concerned extending life. Tompkins claims that TRW made important breakthroughs in developing pharmaceutical products that could extend life through the principle of age-regression. In *Insiders Reveal Secret Space Programs and Extraterrestrial Life*, one chapter discusses the testimonies of Corey Goode, Randy Cramer and Michael Relfe, who all claim to have experienced age-regression technology as part of their respective "20 and back" service with secret space programs. Tompkins has since confirmed that age-regression is a standard practice for many personnel in the "20 and back" programs.[200]

Support for Tompkins' testimony comes from geneticists who have recently publicly identified the genes that control the aging process. In stunning experiments, the results of which have been released in peer reviewed scientific journals, geneticists have demonstrated that they were able to reverse the aging process to varying degrees of success. The lead genetic scientist in the publicly announced age reversal studies is Dr. David Sinclair, who discussed in a November 2014 interview the results of his genetic experiments first conducted on mice:

> We've discovered genes that control how the body fights against ageing and these genes, if you turn them on just the right way, they can have very powerful effects, even reversing ageing – at least in mice so far... We fed them a molecule that's called NMN and this reversed ageing

> completely within just a week of treatment in the muscle, and now we're looking to reverse all aspects of ageing if possible.[201]

Yet, another "Special Project" was a contact program where personnel could meet and interact with Nordic extraterrestrials at a remote mountain location in the Cascades. Tompkins has described how holographic technologies were used to help participants learn about the history of extraterrestrial life, and their interactions with Earth. It is worth repeating that according to Tompkins, Nordic extraterrestrials were covertly helping the Navy in a project that would facilitate open cooperation between them in the future.

Corroborating Tompkins' Testimony & Documents

Tompkins has privately supplied a November 1995 list of Navy, Marine, and Air Force officers, along with local officials who participated in the Navy League, Medford Council. From this list, I was able to locate two Navy officers. In a phone conversation, each confirmed serving with Tompkins. Both of their names also appear in the October 1993 "Minutes of Meeting" document presented earlier.

First, I spoke with Art Lumley (Commander, USN ret.) on May 19, 2016, who at the time was the serving President of the Rogue Valley Council Navy League.[202] Lumley served for 20 years with the Navy and Navy Reserve, and then worked for 26 years as a pilot with United Airlines, retiring as a Captain. He recalled that extraterrestrial life was a topic discussed in "Special Projects" as areas of interest for the Rogue Valley Council during the time that Tompkins was President. He said that Tompkins would raise the extraterrestrial topic in discussions, and it was known among council members that he was an expert on the topic. Lumley said he does not recall any direct involvement in such projects himself, but did recall Tompkins' discussion of them.

Next, I talked with Larry Boeck (Captain, USN ret.) on May 24, 2016. His naval service spanned 25 years and he retired as a Captain in the US Naval Reserve in 1995. During his career he commanded five Naval Reserve units. Captain Boeck said that while he does not recall any direct involvement in "Special Projects" himself, he does remember Tompkins talking about extraterrestrial life and giving reports to committee members about them. He also remembered Tompkins discussing hologram technologies as a means of educating Navy League personnel and Sea Cadets. Boeck stated that Tompkins was very knowledgeable about the extraterrestrial topic, and researched it a lot. He said that due to Tompkins' knowledge and contacts, he found him very credible when it came to the topic of extraterrestrial life.

Finally, on November 1, 2016, I spoke by phone with Admiral Marsh too, who confirmed key parts of Tompkins' story. He recalled having had communications with Tompkins about plans to launch a chapter of the Navy League in Medford, Oregon, and that Tompkins had requested permission to visit the naval submarine facility at Bangor, Seattle. Admiral Marsh also remembered meeting with Tompkins in Medford, Oregon, where they discussed the creation of the Navy League, which was a challenging organizational process. He said he vividly recalled Tompkins' interest and knowledge about earthquakes and this topic's relevance for Naval operations.

I asked Admiral Marsh why he had decided to attend the opening of the Navy League Council in Medford, and he said that it was regarded as an important event for the Navy League, which prior had no official presence in that region of Oregon. At least in part, this corroborates Tompkins belief that it was indeed unusual for an active admiral to launch a Navy League Council.

Tompkins says that he met with Admiral Marsh a second time, this time in Seattle, where they discussed "Special Projects" involving UFOs and extraterrestrial life. I asked the Admiral whether he recalled this second meeting and the topics Tompkins said were discussed. Admiral Marsh stated that he didn't recall the second meeting, but included that he was not surprised that Tompkins was interested in UFOs and extraterrestrial life given his wide range of interests. The admiral added that he had no direct knowledge of the subject of extraterrestrial life. Then he further explained that due to his heavy schedule with many daily meetings with many members of the general public, it was very possible he did meet with Tompkins in Seattle even though he didn't recall it. However, he did recall that Tompkins and his Navy League colleagues did complete the submarine ride at the Submarine Group Nine facility at Bangor, which he was in charge of at the time.

Conclusion

With regard to the Navy League in Oregon being involved in a number of extraterrestrial related "Special Projects" from at least 1991 to 1999, Tompkins has provided documents that support his key claims. The most significant document is the 1993 "Minutes" of a meeting discussing reports by a "Special Projects Committee," and the names of officials attending that meeting.

Critically, two officials of the Navy League at the time, who are identified on the "Minutes" document, have confirmed their participation in the Rogue Valley Council meetings and Tompkins' reports and discussions concerning extraterrestrial related projects. Both Captain Boeck and Commander/Captain Lumley acknowledged that Tompkins was viewed by members of the Rogue Valley Council as an expert on the extraterrestrial issue and advanced aerospace technologies. In addition, Admiral Marsh has confirmed that Tompkins was regarded quite highly by senior figures in the Navy League at the national level, and that Tompkins was very knowledgeable about topics of interest to the Navy.[203]

What adds a further layer of credibility to Tompkins' claims is the testimony of Catherine Austin Fitts, a former Assistant Secretary of the Department of Housing and Urban Development. She says that in 1998, she was told that the Navy had initiated a plan to adjust its operations for a future where it is openly known that extraterrestrials exist and live among us. Fitts was asked to participate in strategy sessions that were being conducted by the Arlington Institute, a non-profit organization headed by John Peterson, which was assigned a number of Navy contracts. Back in September 2002, Fitts wrote:

> In 1998, I was approached by John Peterson, head of the Arlington Institute, a small high quality military think tank in Washington, DC... John asked me to help him with a high level strategic plan Arlington was planning to undertake for the Undersecretary of the Navy.... I met with a group of high level people in the military in the process --- including the Undersecretary. According to John, the purpose of the plan ... was to help the Navy adjust their operations for a world in which it was commonly known that aliens exist and live among us.[204]

In support of her testimony, Fitts later released the "Minutes" of a March 26, 1999 Board of Directors meeting which confirm that discussions of extraterrestrial life were being conducted by the Arlington Institute, as at least one of its contracts—very likely with the Navy— required.[205]

It is more than coincidental that while Tompkins was conducting extraterrestrial "Special Projects" out of the Rogue Valley Council of the Navy League from 1991 to 1999, aiming to educate Navy personnel and their children about extraterrestrial life, that the Arlington Institute was doing something very similar on the other side of the country in 1998. In both cases, senior Navy officials had authorized these initiatives through private civilian organizations.

Consequently, documents and the supporting testimony of two Navy officers confirm that from 1991 to 1999, the Medford Council of the Navy League did discuss extraterrestrial related activities that were occurring in "Special Projects." A third Navy officer, Admiral Marsh, corroborates Tompkins' expertise and connections in a variety of more conventional topics, e.g., earthquake studies, which he thought might be connected to the extraterrestrial topic. Marsh's command of Submarine Group Nine at Bangor, and the permission he gave to Tompkins and his colleagues to be provided a tour of a submarine while in operation, does raise the possibility that this was done to aid one of Tompkins' "Special Projects."

All this supports Tompkins' claims that he had set up a number of "Special Projects" in Medford, Oregon, aimed at educating Navy and other military personnel, along with their children enrolled in the Sea Cadets, about extraterrestrial life and technology, and training for a secret Navy space program. In doing so, Tompkins had the direct support of senior leaders in the national organization of the Navy League. More significantly, he had the support of high level officials in the Department of the Navy who approved for a busy Rear Admiral, holding two command positions, to attend the 1991 opening ceremony of the Rogue Valley Navy League Council.

Endnotes

[188] About the Navy League, http://navyleague.org/aboutus/index.html
[189] About the Navy League, http://navyleague.org/aboutus/index.html
[190] Rense Radio Interview with William Tompkins with Maj. George Filer & Frank Chille – May 4, 2016. Transcript available at: http://spherebeingalliance.com/blog/the-amazing-story-continues-part1.html
[191] Wikipedia, "USS Nautilus (SSN-571)," https://en.wikipedia.org/wiki/USS_Nautilus_(SSN-571).
[192] Phone interview with Bill Tompkins on December 2, 2016.
[193] The Solar Warden Space Program is discussed in depth in *Insiders Reveal Secret Space Programs and Extraterrestrial Life* (Exopolitics Institute, 2015) 147-78.
[194] Phone interview with Bill Tompkins on December 2, 2016.
[195] Commander, Operational Test and Evaluation Force, http://www.public.navy.mil/cotf/Pages/history.aspx A brief biography of Rear Admiral Webster is available at: http://cg22.ussengland.org/newsletters/julyaugust_2001.htm
[196] Commander, Logistics Group Western Pacific, http://www.clwp.navy.mil/History/
[197] William Tompkins, *Selected by Extraterrestrials* (Createspace, 2015) backcover.
[198] William Tompkins, *Selected by Extraterrestrials*, p. 431.
[199] Unpublished video recorded by Michael Salla, March 2016
[200] Transcript of interview on Gaia TV, September 25, 2016, http://spherebeingalliance.com/blog/transcript-cosmic-disclosure-validating-the-20-and-back-program-with-william-tompkins.html
[201] Sue Lannin, "Scientists reverse ageing process in mice; early human trials showing 'promising results,'" http://www.abc.net.au/news/2014-11-04/scientists-reverse-ageing-process-in-mice/5865714
[202] Private phone conversation with Art Lumley, May 19, 2016,
[203] Notes from private phone conversation with Admiral Marsh, November 1, 2016.
[204] See Catherine Austin Fitts, "UQ Wire: What's Up With the Black Budget?" http://www.scoop.co.nz/stories/HL0209/S00126.htm
[205] See Michael Salla, http://exopolitics.org/us-navy-plan-to-disclose-extraterrestrial-contact-secretly-developed-in-1998/ The Arlington Institute document is available here:

---

http://solari.com/blog/docs/2011/Arlington_Institute_Board_Meeting-10Apr2000.pdf

CHAPTER 8

BATTLESTAR GALACTICA & THE ARTIFICIAL INTELLIGENCE THREAT

On March 17, 2009, a select group of United Nations officials learned about war and peace from the perspective of those living on an extraterrestrial mothership with much battle experience against a form of artificial intelligence bent on human genocide. It was not however real-life extraterrestrial veterans imparting their knowledge of war and peace in space, but the creators and cast members from the popular TV series, *Battlestar Galactica*.

Special significance marks this United Nations Panel event because the *Battlestar Galactica* series was created by Glenn Larson, another protégé of Leslie Stevens IV. In chapter five, I presented elements surrounding the unusual business relationship between Stevens and Gene Rodenberry, and the conclusion that Rodenberry was told about a secret Navy Space Program. Due to Stevens' military intelligence background and briefings he received from his father, Vice Admiral Leslie Stevens III, he was privy to classified information that became a key inspirational force behind Rodenberry's *Star Trek*.

Stevens had a similar business relationship with Larson, and in fact wrote the original script for the *Battlestar Galactica* pilot episode. This raises the intriguing possibility that key elements of *Battlestar Galactica* are based on real life threats in space as perceived by senior Navy officials running a secret space program. As a prelude to examining this space threat, it is first worth learning what the United Nations thought it would gain from its Battlestar Galactica Panel.

The United Nations *Department of Public Information* (DPI) and the Sci Fi channel co-hosted the panel, which was scheduled a day before the closing finale to the four season series. The panel explored "themes of importance to both the United Nations and Battlestar Galactica" as described by a March 16, 2009 United Nations Press Release:

> UNITED NATIONS PUBLIC INFORMATION DEPARTMENT, SCI FI CHANNEL TO CO-HOST PANEL WITH BATTLESTAR GALACTICA CREATORS TO RAISE PROFILE OF HUMANITARIAN CONCERNS
>
> ... The discussion will explore some of the themes that are of importance to both the United Nations and the critically acclaimed television show: human rights; terrorism; children and armed conflict; and reconciliation and dialogue among civilizations and faiths.[206]

The Press Release went on to mention the panel participants and that it would be moderated by Whoopi Goldberg—who had previously starred as a popular character in Rodenberry's *Star Trek: The Next Generation*:

> The panel will be moderated by Academy Award-winning actress and producer Whoopi Goldberg. Oscar-nominated actress Mary McDonnell, Emmy Award-winning and Oscar-nominated actor Edward James Olmos, and *Battlestar Galactica* creators and executive producers Ronald D. Moore and David Eick will participate in the panel.
>
> Also participating will be Radhika Coomaraswamy, Special Representative of the Secretary-General for Children and

> Armed Conflict; Craig Mokhiber, Deputy Director, New York Office, Office of the High Commissioner for Human Rights; Robert Orr, Assistant Secretary-General for Policy Planning, Executive Office of the Secretary-General; and Famatta Rose Osode, Minister and Deputy Permanent Representative, Permanent Mission of Liberia to the United Nations.

The Press Release explained what it was about the *Battlestar Galactica* series that was of special relevance to the United Nations:

> "As one of the launch projects of the Creative Community Outreach Initiative, this event will show how skillful storytelling can elevate the profile of critical humanitarian issues," said Kiyo Akasaka, United Nations Under-Secretary-General for Communications and Public Information. "Not only does it present an opportunity for creative discussion, but, more importantly, it offers a chance to deliver a message about the many harsh realities that still exist worldwide."

The Press Release then described some of the topical content of the *Battlestar Galactica* series and why this contributed to it winning numerous awards and accolades:

> *Battlestar Galactica* is the gripping saga of humanity's last remnants and their struggle to find a new home while fleeing from their deadly Cylon enemies. Its issues-driven topicality and command performances have garnered it numerous awards, including several Emmy Awards and the prestigious George Foster Peabody Award.[207]

Shortly after the 2009 event at the United Nations, I contacted Juan Carlos Brandt, Chief of Advocacy and Special Events at the UN, who explained that the initiative grew out of the Sci Fi channel's interest in participating in the Secretary-General's

Creative Outreach Initiative. [208] The initiative was launched in July 2008 by Secretary-General Ban Ki Moon and aimed to foster collaboration between the entertainment industry and the UN.

> "They came to us and explained that there were themes common to both the show and the UN" according to Mr. Brandt, and "that those themes could be discussed here in a serious manner."[209]

He explained that this would be the first time that a TV show has been invited to participate in a UN panel discussion. The choice of *Battlestar Galactica* to cast light on global conflicts was intended to seriously explore the virtues of reconciling cultures with a history of armed conflict.[210]

Summarizing the plot, the show depicts 50 thousand human survivors from a devastating war in a double binary star system, which had 12 colonies on planets with a combined population of 50 billion. The surviving humans escape aboard spaceships and are pursued by a form of Artificial Intelligence (AI), called Cylons, bent on the genocidal destruction of all human life.

Prior to the outbreak of the "Second Cylon War" (the first had ended in a stalemate 40 years earlier) the Cylons learned to replicate human bodies controlled by AI to create humanoid Cylons. The Cylons proceeded to infiltrate the extraterrestrial human society on the twelve colonies before launching another all-out war to wipe out humanity. After all but destroying humanity on their twelve colonial planets, the humanoid Cylons again infiltrate the remaining humans seeking to find a legendary 13th colony, which according to ancient legends had left the original home world of the 12 colonies called Kobol.

Kobol had itself reached a high degree of technological development before undergoing a planet-wide war, leading to the exodus of the 13 tribes. 12 left to establish the human colonies in the four star system that became the 12 Colonies of Kobol, while the 13th tribe traveled to a distant world they named "Earth." It turns out as the series develops that the 13th colony are in fact humanoid Cylons, which had left Kobol after an earlier war between humanity and their AI creation. Ironically, the humanoid Cylons on "Earth" underwent a similar conflict process after creating their own mechanical Cylons with AI, which rebelled against them. Five surviving humanoid Cylons from Earth then travelled to the 12 Colonies of Kobol, and ended the First Cylon war by secretly promising to help the rebel mechanical Cylons create humanoid Cylons. The newly created humanoid Cylons decide to launch the Second Cylon War out of vengeance.

By the end of the fourth season of *Battlestar Galactica* in 2009, the fleeing humans and the two generations of humanoid Cylons, have learned the painful lessons of war, and the repetitive nature of human-AI conflict. They are finally ready to seriously embark on a new adventure of reconciliation and dialogue, choosing to integrate the surviving humans and humanoid Cylons into a new hybrid species that would peacefully evolve together in a primitive world. Their new chosen world would also be called "Earth," which turns out to be our planet about 150,000 years ago. The series ends with the revelation that modern humankind are the hybrid descendants of the extraterrestrial human and humanoid Cylon races that merged as a solution to the ancient conflict.

The *Battlestar Galactica* series certainly raised thought provoking themes about war and peace. At least the United Nations Panel who convened to study the series thought so. Therefore I asked Mr. Brandt:

> In what way does the *Battlestar Galactica* depiction of extraterrestrial life practicing reconciliation and dialogue assist humanity in preparations for a future where extraterrestrial life may be discovered?[211]

He responded with a selection from a speech from Asha-Rose Migiro, *Deputy Secretary-General of the United Nations:*

> Nobody can close their doors to cultural intervention. And if we're talking of peace and security, we have to talk of the different cultures that there are, the different beliefs, and how human beings should learn to live with one another. So this is one way that also contributes to peace, contributes to stability and, therefore, creates the conditions for development.[212]

Battlestar Galactica takes themes related to past genocidal conflicts and more recent ethnic cleansing campaigns, to the level of a planet wide war of extermination by cybernetic AI extraterrestrials that have learned to infiltrate and subvert human society. This would certainly not be the form of "cultural intervention" envisaged by Ms. Migiro.

Among the United Nations officials scheduled to attend the Battlestar Galactic panel discussion listed on the Press Release was Dr. Robert Orr, *Assistant Secretary General for Policy Coordination and Strategic Planning*. He earlier served at the US National Security Council, where he was in charge of peacekeeping and humanitarian operations. If the United Nations were to ever be seriously confronted with issues of reconciliation and dialogue with extraterrestrial life and/or AI life forms in conflict with humanity, Dr. Orr would be high on the list of UN officials qualified to lead such an effort.

The choice of *Battlestar Galactica* for discussing themes of war and peace from the perspective of a fictional galactic conflict raises an intriguing question. Was the show's central premise of a galactic conflict between human-like extraterrestrials and an AI life form in any way inspired by real events? In particular, was classified information received by the show's creator, Glen Larson, from Leslie Stevens about a secret Navy run space program in any way related to a global threat posed by an AI life form?

First, it is worth analyzing the relationship between Stevens and Larsons, and the genesis of the "Battlestar Galactica" idea. In an interview with Alan J. Levi, the Director of the original *Battlestar*

Galactica series, Susan J. Paxton made a surprising discovery. In an article titled, "Leslie Stevens is The Creator of *Battlestar Galactica*," she wrote:

> Levi is known to BG fans as the director of "Gun on Ice Planet Zero," but he also directed half of the premiere after Richard Colla was let go by Larson. Levi was a good friend of the late Leslie Stevens, the producer best known for the famous science fiction series The Outer Limits. Recently I interviewed Alan Levi. I had not planned to ask him any questions about the origins of Battlestar Galactica because he had not been involved early enough in the process to know about it. But, out of the blue, with no prompting from me whatsoever, he said, "Well, Leslie Stevens wrote the original script. Leslie was one of my best friends. I do know that Leslie had told me at one time way before he ever got into the script that he had this great idea for a script that he was going to take to Glen Larson and talk about."
>
> In other words, sometime in 1977, Stevens had told Levi about an idea for a series he was going to discuss with Glen Larson, an idea that recognizably was Battlestar Galactica. Now before people start running around screaming that Larson "stole" BG from Stevens, it's clear that whatever happened, Stevens must have agreed with it, though for what reason we cannot at this time know.[213]

In her analysis of the genesis of *Battlestar Galactica*, Paxton pointed out Larson's earlier efforts in developing a sci fi series that had no resemblance to *Battlestar Galactica*. It was really Stevens' idea, who originally pitched it to Larson, according to Levi. Then even more tellingly, Stevens wrote the original script for *Battlestar Galactica*, and was content to let Larson take all the credit.

This is similar to what Stevens had previously done with Rodenberry from 1964 to 1965 while shooting the Outer Limits. A business relationship was developed where they collaborated in setting up the original idea of *Star Trek*. Now, just over a decade later, in 1977, Stevens began doing the same thing with Larson.

It is very unusual for television and movie producers to forego taking credit for their creations, unless there is a deeper agenda at play. In the case of Stevens, his military intelligence connections and involvement with psychological operations led him to disseminate classified information into the movie and television industry to hide the truth in plain sight.

If *Star Trek* was modeled on a real life secret Navy space program cooperating with Nordic extraterrestrials in a galactic battle with Reptilian aliens, what was *Battlestar Galactica* representing? For the answer, we can go to the testimony of secret space program whistleblower, Corey Goode, on the issue of artificial intelligence and extraterrestrial life.

Goode's claims that the space programs he served with, from 1987 to 2007, had identified Artificial Intelligence as an existential threat to humanity. Goode describes elaborate security protocols that have been put into place to detect and eliminate an "AI Signal" that not only has the ability to infiltrate advanced technology, but also biological systems.

> The best way to avoid the dangers of AI is to educate yourself on the potentials they have to cause a loss of sovereignty. Becoming too dependent on technology is something else that will make you more of a target to be controlled by AI influence or even be infected by an "AI Signal" that can live in the bioelectric field of your body. This signal can then have an effect on the way you think and behave. This is something that is screened for even currently when operators and guests arrive at SSP [Secret Space Program] Facilities and is something that other ET Groups take seriously as well.[214]

Goode says extraterrestrial civilizations have much experience in observing how AI has taken over countless other worlds, and led to the eradication of the indigenous population that created the AI. Essentially Goode is warning that the storyline of *Battlestar Galactica* is more genuine galactic history than science fiction:

> Those who are AI Prophets are already working on a timeline to create a society completely dependent on technology that at one point will hand its sovereignty over to this "AI God" because it will be believed that this AI is the only thing that can rule the world from a neutral perspective and bring world peace for the first time. These AI Prophets have been shown the information of the thousands of other civilizations that have fallen for this trickster god model and were all destroyed.[215]

Is there any evidence to support Goode's information that Artificial Intelligence is not only an existential threat to humanity, but has plagued many different extraterrestrial civilizations as depicted in *Battlestar Galactica*?

The Real Life "Friendship" Case of Extraterrestrial-AI Conflict on Earth

A 2009 book titled, *Mass Contacts*, details the history of a mysterious group of human-looking extraterrestrials that established underground bases in Italy, and met with local residents between 1956 and 1978.[216] The author, Stefano Breccia, was a highly regarded Italian UFO researcher with an electrical engineering background, who previously taught at several Italian and foreign universities. He investigated the "Amicizia" (Friendship) case over a period of several decades during which he met and questioned many of the primary witnesses. The most prominent was Bruno Sammaciccia, a prolific Italian author and scholar with many high connections in the Italian national security system, including certain generals and diplomats.

Some of the many photos taken of UFOs and extraterrestrials in the Friendship case rank among the best quality ever taken. Breccia included about 20 of these photos in *Mass Contacts*. In terms of its overall impact, number of participants involved, and documentary evidence compiled, *Mass Contacts* is the most compelling case of direct human extraterrestrial contact in modern history.

Figure 54. Amicizia (Friendship) Flying Saucer Craft Photo from Mass Contact, p. 361

The story of *Mass Contacts* began back in 1956, when Sammaciccia and two of his friends met with two mysterious individuals who said they were extraterrestrials. One was over eight feet tall, while the other was just above three feet in height.

Sammaciccia and his friends, initially skeptical, were eventually taken into a large underground base where they saw more of the alleged extraterrestrials. They also saw their children being educated, some of the advanced technologies they used, and their spaceships. Finally convinced that they were really having physical contact with beings from another world, Sammaciccia and his friends began to help the extraterrestrials. They began with material support by arranging for truckloads of fruit, food and other material to be transported and unloaded at an extraterrestrial base. Eventually, two truckloads of supplies were being delivered every month to bases in different regions of Italy where Sammaciccia and his assistants lived.

Sammaciccia described the various people involved in the case, including those who had direct meetings with the extraterrestrials. The number of individuals involved grew over time, while Sammaciccia continued to assist the extraterrestrials in helping prepare humanity for the reality of human-looking extraterrestrials from other planets. Breccia said that he personally met and interviewed almost 80 people who worked with, or met, the extraterrestrials. Most were in Italy, but others were from other countries and had also been exposed to this same group of extraterrestrials. Many of the extraterrestrials could easily blend into human society and even take normal jobs when necessary. The Friendship case helps confirm other accounts by a number of contactees and whistleblowers who state that human-looking extraterrestrials have blended into human society and have walked among us.[217]

Sammaciccia finally described a violent conflict between two factions of extraterrestrials trying to influence humanity's development and future. While his "Friendship" faction were biologically human and promoted spiritual development, the other faction were advanced humanoid robots with Artificial Intelligence that promoted technological development at all cost. In *Mass Contacts*, Breccia included an interview where one "Friendship extraterrestrial" discussed the history and activities of the robotic life forms known by the acronym, CTR (Italian for "Contrarians"—adversary):

> The CTR are the results of an experiment that has run out of control. They are robots, in the full meaning of this word, even if centuries ago they have started an activity of biological reproduction. To you, at this point, it's no longer possible to discriminate between a natural being and a biological robot....[218]

The alleged extraterrestrial elaborated further on how these synthetic AI robots were indistinguishable from organic humans:

> The CTR's (and, by the way, not them only) are on the contrary artificial creatures, and they remain as such, even if nobody among your physicians would be able to discriminate; actually there is nothing physical that may allow to state whether an organism is a descendent from artificial robots or not.[219]

The philosophical and biological differences between the two extraterrestrial factions in the Friendship case led to periodically violent clashes. Eventually, the underground bases of Sammaciccia's extraterrestrial friends were destroyed in 1978. This included their largest base, which was under the Adriatic Sea, just off the Italian coast. Survivors had to leave the Earth, but promised to return later when humanity was ready for a more ethical future when it could openly interact with extraterrestrials.

The story revealed by Breccia is astoundingly similar to the storyline of *Battlestar Galactica*, where the humanoid Cylons have evolved to the extent that they are physically indistinguishable from biological humans and launch a genocidal war against their human creators. The humanoid Cylons chase the humans all across the galaxy until they reach the Earth and settle here. The humanoid Cylons have evolved so far as to be capable of physical reproduction and interbreeding with humans, which is how the series ends with a new hybrid Human-Cylon race populating the Earth 150,000 years ago, to create the current human civilization.

Battlestar Galactica has many elements that correspond to information released by Corey Goode and Stefano Breccia about AI life forms being an existential threat to human civilizations throughout the galaxy. It is markedly significant that the alleged defeat of the Friendship extraterrestrials by their AI (CTR) humanoid robot opponents took place in 1978, which coincides with the time Stevens worked with Larson, resulting in the launch of the *Battlestar Galactica* series. Was this mere coincidence, or was Stevens passing on classified information about the real and present threat posed by AI humanoid extraterrestrials to humanity?

It's very likely that NATO and the US Navy in particular, closely monitored the escalating conflict between the two extraterrestrial factions, and the resulting violent destruction of the bases belonging to the more peaceful human-looking extraterrestrials by the AI humanoid group. After all, Italy was a NATO ally at the time, and everything Sammaciccia was doing to help the human-looking extraterrestrials was being closely monitored by Italian national security authorities, and tacitly allowed by them. Destruction of the extraterrestrial bases, knowing fully well the surrounding circumstances, would certainly have set off alarm bells for the US Navy, which was monitoring ocean activities through its naval bases in Italy at the time.[220] The violent destruction of underwater bases would have set in motion a process whereby Leslie Stevens, due to his military intelligence background and continuing intelligence mission with Hollywood, was given a classified briefing of what was happening and the threat posed by AI life forms.

The 2009 United Nations Panel on Battlestar Galactica emphasized the importance of the reconciliation and conflict resolution depicted in the series finale, especially when it came to resolving complex ethnic and social problems. Perhaps this was a way for the United Nations to encourage one solution to a complex galactic conflict between organic and AI life forms, which was now involving our planet.

In contrast, Goode emphasized the danger of infiltration by an AI signal that could penetrate both advanced technology and human biology. He says that technology has been developed that can distinguish between humans and AI life forms, and is used as a security measure by different secret space programs and extraterrestrial civilizations. According to Goode, some leaders of the Draconian Reptilians had been compromised by this AI signal in the form of nanites: "it turns out that a lot of these reptilian higher castes are highly infected with nanites—artificial intelligence-type of nanites."[221]

The threat posed by AI is consistent with the original message found in the first *Battlestar Galactica* series that was influenced by Stevens, and tacitly by leaders of the Navy space program, then making final preparations for its secret launch in the early 1980's, while also monitoring developments in Italy. This suggests great caution in how humanity should deal with the complex social, legal and national security issues posed by the development of AI, and the possibility that AI life may be attempting, or has already succeeded in, efforts to infiltrate human society, secret space programs and even some visiting extraterrestrial groups.

Endnotes

---

[206] United Nations Press Release available online at: http://www.un.org/press/en/2009/note6192.doc.htm
[207] United Nations Press Release available online at: http://www.un.org/press/en/2009/note6192.doc.htm
[208] Quoted in Michael Salla, "United Nations Panel Discusses Extraterrestrial War and Peace," *The Examiner*, http://www.bibliotecapleyades.net/exopolitica/exopolitics_UFOUN08.htm
[209] Quoted in Michael Salla, "United Nations Panel Discusses Extraterrestrial War and Peace," *The Examiner*, http://www.bibliotecapleyades.net/exopolitica/exopolitics_UFOUN08.htm
[210] United Nations Press Release available online at: http://www.un.org/press/en/2009/note6192.doc.htm
[211] Quoted in Michael Salla, "United Nations Panel Discusses Extraterrestrial War and Peace," *The Examiner*, http://www.bibliotecapleyades.net/exopolitica/exopolitics_UFOUN08.htm
[212] Quoted in Michael Salla, "United Nations Panel Discusses Extraterrestrial War and Peace," *The Examiner*, http://www.bibliotecapleyades.net/exopolitica/exopolitics_UFOUN08.htm
[213] Susan J. Paxton, "Leslie Stevens is The Creator of "Battlestar Galactica," http://languatron1.blogspot.com/2011/03/leslie-stevens-is-creator-of-battlestar.html
[214] "Questions for Corey Goode on SSP Conflicts and Human Slave Trade," http://exopolitics.org/galactic-human-slave-trade-ai-threat-to-end-with-full-disclosure-of-et-life/ (accessed 8/10/15).
[215] "Questions for Corey Goode on SSP Conflicts and Human Slave Trade," http://exopolitics.org/galactic-human-slave-trade-ai-threat-to-end-with-full-disclosure-of-et-life/ (accessed 8/10/15).
[216] Stefano Breccia, *Mass Contacts* (AuthorHouse, 2009).
[217] Michael Salla, "Extraterrestrials Among Us," *Exopolitics Journal* Vol 1:4 (2006). Available online at: http://exopoliticsjournal.com/vol-1/1-4-Salla.htm
[218] Stefano Breccia, *Mass Contacts*, 272.
[219] Stefano Breccia, *Mass Contacts*, 273.
[220] For list of current US Navy bases in Italy, see: Military Bases in Italy," https://militarybases.com/italy/
[221] Corey Goode interview in "Cosmic Disclosure: The Dark Fleet, Season 4, Episode 6,"

---

http://spherebeingalliance.com/blog/transcript-cosmic-disclosure-the-dark-fleet.html

CHAPTER 9

THE CURRENT OPERATIONS OF SOLAR WARDEN

Thus far, William Tompkins' role in the design and development of a Navy Secret Space Program during his almost 40 year Navy and aerospace career has been laid out at length. The Navy's space program, which Tompkins says was called Solar Warden since its inception, became operational in the early 1980's. With his retirement from full time employment in the aerospace industry in 1984, Tompkins devoted himself to a new chapter in his life where he would influence the Navy's policy on how to prepare humanity for the truth about extraterrestrial life and the advanced technologies used in its space program. This policy involvement would be done through his membership in the Navy League in Oregon (1991-1999), and more recent active participation in the Navy's annual "West" meetings in San Diego (2000-2016).

Tompkins' knowledge about the Navy's Solar Warden program goes beyond a detailed understanding of its origin and development stages, and remarkably extends to current operations. This is highly significant since many of his revelations help corroborate the testimony of others who have also come forward to reveal operational details of Solar Warden. Of these, Corey Goode's claims are the most significant in terms of their scope and similarities to those of Tompkins.

In short, Goode says that he was covertly recruited into Solar Warden in 1987, and served in it and several other secret space programs until 2007. During this time, he had access to "smart glass pads" which contained extensive briefing documents about the history, development, and operations of Solar Warden and other secret space programs. Goode recalls that some of the documents he read on the smart glass pads looked like they had had been prepared decades ago by stenographers.

When Goode's revelations were first released in book form by me in *Insiders Reveal Secret Space Programs and Extraterrestrial Life* (September 2015), neither Tompkins nor Goode knew of each other. It was only *after* the publication of Tompkins book, *Selected by Extraterrestrials* (December 2015) that its editor, Dr. Robert Wood, received a copy of my book. After reading it, he noted the remarkable similarities between my analysis of the history and development of secret space programs, which was based largely on Goode's testimony, and what Tompkins was saying in his own newly released book.

After further investigation, it turned out that the substantive match between Tompkins and Goode's respective histories of the secret space programs could be traced back to the 29 Navy spies and approximately 1200 nightly debriefings that Tompkins attended from 1942 to 1946. These debriefings run by Admiral Rico Botta out of Naval Air Station, San Diego were recorded by stenographers. In addition, the briefing packets prepared by Tompkins and delivered by him to various aerospace companies, think tanks and research institutions, were assembled with the assistance of a pool of typists and copy girls.

Apparently in the due course of maintaining historical archives, both the records of the Navy spies' debriefings and Tompkins' subsequent briefing packets from the World War II period were uploaded onto the smart glass pad's expansive database. Rather fatefully, more than forty years later, Goode would read these same documents that Tompkins had prepared or was privy to when stationed at Naval Air Station, San Diego. Goode later told in an interview how he came to the realization that the briefing documents Tompkins was discussing were the same ones he had read:

> I'm really beginning to believe that a lot of his briefings during the time period – 1942 – were what made it into the database that I was reading on the smart glass pad, you know, like 30 [40] years ago... we were viewing old typeset documents.[222]

Tompkins has described his involvement in two space programs.[223] First came his assignment in naval intelligence operations on a German/Nazi led secret space program during World War II. Tompkins learned that the most advanced elements of this program were relocated to Antarctica after a secret agreement was reached with an extraterrestrial Reptilian race, called the Draconians. This German/Nazi space program in Antarctica evolved into what Goode later described as the "Dark Fleet," which would essentially form a mercenary space force to assist the Draconians in their plans of galactic conquest.[224]

Many key elements in Tompkins' revelations about the subsequent activity of the German/Nazi space program correspond to what Goode has independently released from his sources. The Nazi/German space program used German scientists that were part of Project Paperclip to infiltrate the US space programs and military industrial complex. This strategic action, together with what both Tompkins and Goode described as a Nazi fly over of Washington DC in the summer of 1952, led to subversive agreements with the Eisenhower administration. Goode says these agreements were a "silent coup":

> During the 1950's and after they [NAZI's] had successfully infiltrated and subverted the Military Industrial Complex and major Corporate heads they had effectively won control of the direction of not only the Break Away Civilization

> Programs but also the mainstream government and financial system. It was a very effective and silent coup that gutted what was once the American Republic and turned it too into a Corporate Entity with each of us being "Assets" with our very own serial numbers.[225]

Both Tompkins and Goode say that the Navy played catch-up for several decades with the more advanced German/Nazi space program. It was only in the early 1980's that Solar Warden became operational with its first spaceship deployments in what would ultimately lead to eight space battle groups. Tompkins asserts that the key to the Navy finally succeeding in its efforts was the covert assistance provided by Nordic extraterrestrials, who offered highly advanced engineering expertise to help design and build the future space battle groups, despite Nazi-Reptilian sabotage efforts.

Tompkins' Sources for Current Solar Warden Operations

Knowledge about the current operations of secret space programs derive from both terrestrial and non-terrestrial sources for Tompkins, and are independent to those of Goode. Foremost, Tompkins says he continues to be linked to the human-looking Nordic extraterrestrials. In fact, he says they are the same ones who formally assisted him when he was working for various corporations in developing designs for many different classes of spacecraft that would enter into either: a. the Apollo Program; b. the Nova Program; or c. the Solar Warden program.

Tompkins explains that he would be given "flashes" or "downloads" of information revealing complex details about the designs he was working on. These extraterrestrial-assisted designs would be passed on and Tompkins emphasizes that they invariably worked the first time. This was a communication process that

Tompkins' superiors, both at Naval Intelligence and Douglas Aircraft Company, were familiar with. Tompkins recalls what Elmer Wheaton, the head of the Advanced Design think tank at Douglas, told him back then:

> An extraterrestrial race has contacted our government, and from what we have learned they be Nordic people; very similar to us. Navy Intelligence recruited you because they knew that you and others like you were visited and selected as children, and now, in your adult lives, these entities have been communicating with you telepathically.[226]

Wheaton, according to Tompkins, went on to state that the Office of Naval Intelligence viewed Tompkins as vital for distinguishing between different groups of extraterrestrials that were covertly intervening in human affairs:

> Naval Intelligence has informed us of your capabilities and that your judgement is impeccable. As a contactee, you've been put in a position where you can begin to sort out why these different races are at war with each other. More important, what our involvement on this planet is and what should be.... You've been implanted with a program that transfers highly advanced information from extraterrestrials to what the Navy calls 'preferred human contactees.'[227]

How can it be verified that the Navy regarded Tompkins as a "preferred human contactee" with a unique telepathic connection to friendly Nordic looking extraterrestials? In chapter seven, I introduced different types of evidence that Tompkins, with the active support of senior Navy officers, was allowed to establish "Special Projects" for the Navy League Rogue Valley Council that involved educating retired/reserve Navy officers and their children about extraterrestrial life. The documents and testimonies of the retired Navy officers, confirming key aspects of Tompkins' claims, offers meaningful support for the accuracy of what Tompkins recalls from his conversation with Wheaton.

Tompkins says he continues to receive telepathic flashes/downloads, which reveal current operational details about both extraterrestrial and Navy vehicles in space. He believes this is because the Nordic extraterrestrials have formed a covert alliance with the Navy's Solar Warden space program. This means that Tompkins continues to be an important intermediary for this alliance, which operates covertly due to the formal agreements between different US administrations and the Nazi-Reptilian alliance, as detailed in chapter three.

The second source for Tompkins' knowledge about current Solar Warden operations comes from his time at the helm of the Rogue Valley (Medford, Oregon) chapter of the Navy League from 1991 to 1999. During this period, he was given briefings about operations of secret space programs and extraterrestrial life, and shared some of this information through a "Special Projects Committee" with retired Navy, Marine and USAF officers and their children.

Tompkins says that Admiral Hugh Webster, who ran the Navy League chapter in San Diego and more importantly was an executive officer for it nationally, was very familiar with the Navy's secret space program operations and the historic role played by Admiral Rico Botta in the program. He was also very aware of Tompkins' key role as a "preferred human contactee" possessing reliable information from a friendly group of extraterrestrials covertly assisting the Navy. In 2000, when Tompkins moved to San Diego, he had more opportunities to work directly with Webster. It was Admiral Webster who gave Tompkins permission to reveal in his autobiography all that he knew about the Navy's secret space program, and the covert alliance with Nordic extraterrestrials.[228]

Furthermore, Tompkins has disclosed that he is a regular participant in the Navy's annual "West" meetings in San Diego, where leading aerospace companies come together to share their research and development findings. The annual West meetings are organized by the Armed Forces Communications and Electronics Association (AFCEA), describing itself as follows:

> AFCEA provides a forum for military, government and industry communities to collaborate so that technology and strategy align with the needs of those who serve. We are a member-based, non-profit international organization that has helped members advance information technology, communications and electronics capabilities since 1946.[229]

Among the many purposes of these annual "West" meetings is the goal to learn about the latest technologies that may be useful for the Navy's secret space program. Tompkins reports that he has attended all Navy West meetings since the year 2000. In February 2016, he attended again and says he was given access to classified meetings, where plans to disclose Solar Warden to the general public was discussed. Advanced technologies such as "life extension," according to Tompkins, are scheduled to be released in the next two years as part of the Navy's disclosure initiative.

What makes Tompkins' claims very plausible are recent announcements, by biologists such as Dr. David Sinclair, that life extension technologies are scientifically feasible based on open source studies on mice. Dr. Sinclair explained how the life extension process could also be done safely for humans:

> We've gone from mice into early human studies actually. There have been some clinical trials around the world, and we're hoping in the next few years to know if this will actually work in people as well ... They show that the molecules that extend lifespan in mice are safe in people.[230]

In a November 2014 interview, Dr. Sinclair went on to say that drugs based on the nicotinamide mononucleotide (NMN) molecule could be successfully developed "to restore youthfulness in human cells."[231] Sinclair's view that NMN based drugs will eventually be developed for safe use by humans is exciting and solid corroboration for Tompkins' claims.

In private, Tompkins has shown both Dr. Robert Wood and I the pass he used to gain access for the West 2016 meetings. This important piece of evidence that we witnessed makes it very possible that the events he described did occur. It also confirms that senior Navy officials still view Tompkins as very important for their long term policies concerning Solar Warden, no doubt due to his long historical background in developing it and his ongoing communications with Nordic extraterrestrials that still continue today.

Tompkins' testimony about what he learned at the 2016 Navy West meetings correspond closely with Goode's information from ongoing sources. The Solar Warden program, in Goode's account, is linked to an alliance of space programs called the Secret Space Program Alliance, which has been developing strategies for "full disclosure."[232] This would involve the disclosure of: multiple secret space programs; the existence of extraterrestrial life and of ancient human civilizations in the Earth's interior; and finally the release of many advanced technologies that would revolutionize life on our planet. Goode has been reporting that since March 2015, up to the present, there have been sustained negotiations over how much of the truth should be disclosed to the general public.[233]

At the time of this writing, Tompkins is planning to attend the Navy West conference in February 2017, and crucially, he still has the necessary security clearance to be present at classified meetings. Tompkins is currently 93 years old, with a sharp mind, excellent memory and he is still physically healthy. It would be fair to say that he is considered to be a "wise elder" among those running current operations within the Navy's secret space program, and highly respected for continuing to be a liaison with Nordic extraterrestrials.

What is Happening in Solar Warden Today?

Given Tompkins' background, and continuing involvement with the Navy's Solar Warden program, his statements about its current operations bear special significance. In a September 2016 interview, he publicly revealed important information about it:

> Thousands, thousands, not just a few, thousands of people have joined the Navy here in the United States. They joined the Space Navy. They signed up for a 20-year tour. So these folks, men and women, were given a lot of examinations and a lot of information on what they were going to need.
>
> Many of them went to the Moon, our Moon, and facilities there and got checked out and organized and established where is the best place they're going to go, what their criteria is going to be, what major area they're going to develop, like what class in the university.[234]

Tompkins' comments here corroborate the testimony of Goode, who also asserts that there are thousands of personnel involved in the Navy's secret space program. Goode, and another alleged whistleblower, Randy Cramer, spoke of an operations center on the Moon called "Lunar Operations Command," and both claim they were taken there in 1987 for their formal inductions.[235]

How secret space program personnel are assigned after their induction process is described by Tompkins, who says:

> And then they get assigned to a specific base. And they work for a short time in the base before they're assigned a naval cruiser, or a naval attack vehicle, or even a naval spacecraft carrier, which are one, two, and four kilometer long class. And we have eight of those battle groups out there. So there's plenty of room for new people to come aboard....[236]

The above statement corroborates Goode's description of what he personally experienced after his induction at "Lunar Operations Command." He says he was assigned to a research vessel, the Arnold Sommerfeld, where he spent six years performing a number of science and exploratory missions in the solar system.[237]

Tompkins next corroborates Goode and Cramer's claims, along with a third whistleblower, Michael Relfe, about what happened at the end of their 20 and back programs. Goode and Cramer say they were age-regressed 20 years and respectively sent back in time to 1986/1987, and at that time their military memories were wiped. Relfe said that he had the same thing happen to him as he was returned to 1976, to then relive the next twenty years of his life again as a young man. Tompkins says:

> So at the end of the 20 years, they have an option for another 20 years. They could go for another 20 years. Or they have the option to come back to Earth where they were born and where they entered the Navy. And then they make this decision they want to come back. So they age regress. They're now 20 years older than when they joined the Space Navy, okay? And so they take several weeks or several months, and they reverse their age back to 21 when they joined the Navy.... And now, during that several weeks returning, their minds, not painfully, but their minds are played with to where 90.99 [likely meant to say 99.99 percent] of their memory for the last 20 years out in space is removed.... that system is operating and has been operating since 1980.[238]

In *Insiders Reveal Secret Space Programs and Extraterrestrial Alliances*, a chapter is devoted to a comparative study of the testimonies of Goode, Cramer and Relfe.[239] Tompkins' testimony adds a significant level of credibility to what each of these individuals had to say about the process they underwent at the end of their 20 year tours of duty in their respective secret space programs.

Tompkins has further revealed his knowledge about a corporate run space program that emerged to rival the Navy's Solar Warden program:

> But the corporations, at the same time they were doing the military mission programs, they were doing other programs which could possibly allow them to nullify the space programs of the military, like Solar Warden. And these people, then, utilizing all the advances in space systems, militarily wise, developed the capability to move off the planet and mine materials on other planets in the solar system and/or continue on out into the galaxy through the 12 closest stars, Alpha Centauri being the first, and mining or extracting materials, or whatever, and making money.
>
> This same group of top corporate people of all these companies that are doing the jobs, that they are getting paid for, were paralleling the space missions for industrialization – making money... We've got corporate operations operating parallel to missions to solar system planets and other stars' planets, doing the same thing.[240]

This testimony by Tompkins corroborates what Goode revealed in early 2015 about the existence of a corporate run space program; Goode specifically calls it the Interplanetary Corporate Conglomerate.[241] Back in April 2014, Cramer claims he was assigned to a Mars military base from 1987 to 2004, to protect five corporate run colonies that he identified as the Mars Colony Corporation.[242] Similarly, Relfe says that during his time on Mars, from 1976 to 1996, he was part of a military operation to protect Mars colonies.[243]

Based on Tompkins' identification of the corporations involved in the construction of the Navy's secret space program at underground facilities in Utah, we know which of them are very likely to form the core of a similar corporate run program—the "Interplanetary Corporate Conglomerate." Previous material in chapter six includes Tompkins naming Lockheed Martin, Northrup Grumman and Boeing Aerospace as the major corporate contractors responsible for building the Navy's kilometers-long spacecraft. It was pointed out how each of these three companies could be directly linked to predecessors who had received the briefing packets Tompkins had put together from Navy spies' debriefings.

As primary contractors for building a corporate run space fleet, these corporations would sub-contract out to hundreds of other companies, using basic compartmentalization security procedures which are standard for Unacknowledged Special Access Programs.[244] The various components for the kilometers-long spacecraft, as described by Tompkins, would be assembled together using modular construction practices. Modular construction, which was first used decades ago in Unacknowledged SAP's, has recently been declassified for modern engineering use in conventional aerospace and maritime projects.

Here is how General Dynamics/Electric Boat, for example, describes the use of modular construction for its new Virginia-class of nuclear powered submarines.

> Working closely with the Navy and its industry partners, Electric Boat agreed to reduce the cost for Virginia-class submarines to $2 billion per hull in Fiscal Year 2005 dollars. This successful effort comprises three parts: multi-year procurement, improvements in construction practices and design for affordability...
>
> The Virginia Class has been designed and built using advanced electronic design and data management tools that have been integrated into modular construction techniques.

The use of Integrated Product and Process Design (IPPD) is credited with contributing to the success of the program.

Improvements in construction performance will reduce construction span from 84 months to 60 months. This is being achieved through greater use of modular construction, pushing as much work as possible into a manufacturing setting where it can be done more efficiently. USS New Hampshire was the first ship to be assembled from four modules, compared with the 10 modules required to build the lead ship of the class, USS Virginia.[245]

Using modular construction practices and compartmentalized security procedures in place for the military industrial complex, large aerospace corporations can build massive kilometers-long spacecraft not only for the Navy, but also for the Interplanetary Corporate Conglomerate, and others.

Figure 55. Virginia Class Submarine built using Modular Construction Practices

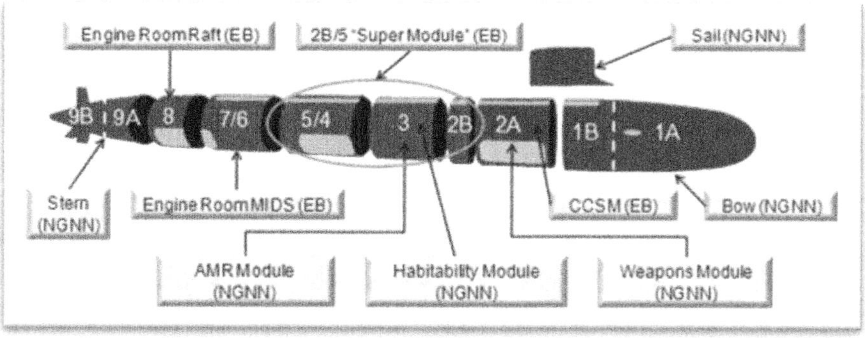

Tompkins' testimony supports Goode's claim that the Interplanetary Corporate Conglomerate essentially became a rogue space program outside of the constitutional authority of the Navy's Solar Warden program. Goode's assertion that the Corporate

Conglomerate became a key ally of the German/Nazi Dark Fleet based in Antarctica, and sabotaged other space programs, is also corroborated by Tompkins. This directly connects to Goode and Tompkins' allegations regarding a slave trade conducted by the Dark Fleet with the active support of the Interplanetary Corporate Conglomerate.

After agreements were reached in 1955, the US military industrial complex began assisting the German/Nazi space program by providing it with resources, and most importantly, personnel. This led to a rapid increase in the use of slave labor by the Dark Fleet, which coordinated closely with its Reptilian allies. On this topic, Tompkins says:

> [T]he Germans were totally more advanced and capable of learning how to operate these vehicles and go out in space with the Reptilian naval groups and operate with them in the galaxy, essentially doing what Germany was going to do to this planet, which was essentially to take out the people they didn't want on the planet and make the rest of them their slaves.[246]

Goode elaborates on the extensive use of slave labor by the Dark Fleet, and the collusion in this by the military industrial complexes of the United States and other major nations.

> The Secret Earth Governments and their Syndicates discovered that a large amount of humans were being taken off the planet by various ET's anyway so they decided to find a way to profit from it and have control over which people were being taken. In prior arrangements they were made promises of receiving technologies and biological specimens for allowing groups to abduct humans but the ET's rarely delivered on their promises. Once they had developed the advanced infrastructure (ICC) in our Sol System along with advanced technologies (that some of the thousands of ET groups traveling through our system were now interested in

> obtaining) and now had the ability to deter most unwelcome guests from entering Earths airspace the Cabal/ICC then decided to use human trafficking as one of their resources in interstellar bartering.[247]

The extensive use of slave labor and a galactic trade in trafficking abducted humans are among the most critical issues being currently negotiated in terms of future "official disclosure(s)," according to Goode. Revealing the true scope of such crimes would predictably lead to demands for members of the Cabal/Illuminati, and complicit global elites, being charged with "crimes against humanity." Predictably, behind the scenes this has led to Cabal/global elite demands for "limited disclosure," where the truth of such crimes is delayed for up to 100 years.[248]

Another issue relevant to current Solar Warden operations concerns giant planet-sized spheres that have recently entered our solar system. Tompkins reveals what he knows of this phenomenon:

> As far as our situation now today, It's been accepted that we have a number of different vehicles, which are actually planets that are hollow, that move through the galaxy to monitor good guys and bad guys, wars and no wars, people taking advantage of other people on different stars' planets, and that, as of about a year and a half ago, one of these parked just outside of the solar system. And it has been there for the past year and a half. And this vehicle is quite large, and it has over 2,000 different extraterrestrial civilization people on board as observers and as monitors.[249]

This information is comparable to Goode's claim that a group of five extraterrestrial races, linked to giant planet-sized spheres (Sphere Being Alliance), have recently entered our solar system in order to monitor and influence the relations between humanity and different visiting extraterrestrial civilizations. According to Goode, this began in the early 1980's and accelerated in 2011, when giant Jupiter-sized spheres entered our solar system.

Next what Tompkins explains in more details is very significant in terms of supporting specific assertions made earlier by Goode. Tompkins claims that the giant sphere(s) that are monitoring our solar system: "have blocked the extraterrestrials who are here, are underground in the caverns, from leaving, and they're not allowing their buddies to come in." [250] This closely matches Goode's claim that the spheres created an "outer barrier" around both the Earth and the entire solar system which has led to a realignment of resources and personnel among different space programs and extraterrestrial visitors:

> There are also some very powerful Secret Earth Government and Syndicate Members/Leaders who have defected and have been granted an "Off World Witness Protection Program" for them and their families for actionable intelligence, evidence against their former leadership as well as promise of being witnesses in future World Courts when there are the "Post Full Disclosure" hearings against all of the groups who are trapped in the Sol System behind the "Outer Barrier".[251]

Tompkins also describes the key functions of the giant planet sized spheres:

> I understand that there's two goals. One of them is to nullify the problem with the Sun. The Sun is alive. Okay? Stars are alive. I think it's a hard time accepting this. And they have moods... If the Sun sneezes, all of our communication – radio, electronics, everything – is affected. So there's this group of people, whatever you want to call them, who are attempting to primarily nullify the effects to a region, which is not just like the solar system or our star's system, but do this as a business so that the levels of catastrophes, things

> that cause dangers to the people and to the planet, are less. [252]

This is very similar to what Goode cites as being one of the key functions of the giant spheres belonging to the Sphere Being Alliance; they are to act as a buffer for the cosmic energies coming into our solar system that initiate enormous changes for all life in our region of the galaxy and trigger the sun into hyperactivity.

> The Sphere Alliance has been mostly focused on the energetic changes occurring in our local star cluster because of the area of the Galaxy they are entering. They have been using the many thousands of cloaked spheres that are equidistantly spread out across our Sol System and neighboring Sol Systems (electrically connected in the "Cosmic Web" and natural portal system) to buffer and diffuse incoming tsunami waves of highly charged energy that change the vibratory state of space/time, energy and matter to cause it to raise to a higher state in the "Density Spectrum". This energetic change has a direct effect on not only every planet and star in the region but also every life form below a certain vibratory level.[253]

The "solar sneeze" Tompkins refers to is much the same as the possibility of the sun's corona being ejected in a massive burst of solar activity as Goode was told.

> The most scientific and pragmatic of these "secret syndicates" believe the entire surface area of the Sun's corona is about to mass eject. They believe the Sun will then go almost completely dark for a number of days before it re-awakens into a new state of equilibrium. These syndicate "egg heads" believe that the proceeding energetic shock wave and then the impact of this Mega CME would take out all communications, electronics and sources of electricity on

> Earth... Some of these syndicates further believe this will cause a flip in polarity of the Earth's magnetic field, having a dramatic effect on every living creature on the planet. The neurology and magnetic fields of our bodies would be affected as would our consciousness.[254]

Tompkins' detailed knowledge of current operations of the Navy's secret space program, a corporate run program, and his willingness to now relate this information is a sizable development for all who have been closely following his testimony. It helps greatly in substantiating the earlier testimony of Goode and others concerning Solar Warden and parallel secret space program operations.

The main contention by Tompkins is that the Navy has been secretly allied with a positive human-looking group of extraterrestrials called the Nordics, and this is especially revealing. It was raised in chapter four that multiple whistleblower and private citizens claim human-looking extraterrestrials have infiltrated all facets of human society, including the military industrial complex of major nations.[255]

Tompkins testimony points to such infiltration having both positive and negative elements. The more negative elements correspond to a Reptilian extraterrestrial race that either directly, or through the use of proxies, has infiltrated the US military industrial complex. The goal has been to co-opt personnel and resources for its mercenary Dark Fleet, or to sabotage the Navy's efforts to establish its own independent space program.

There is also the problem of infiltration by extraterrestrial life forms that use or are dominated by Artificial Intelligence, as discussed in chapter eight. These AI aliens have been combatting more positive extraterrestrial visitors for influence over human affairs. A clear example was illustrated in the case of the Amicizia/Friendship extraterrestrials that were very active in Italy, from 1956 to 1978, before having their bases destroyed by the AI aliens. More disturbingly, these AI aliens have been silently encouraging humanity to develop in ways that make it more dependent on Artificial Intelligence. This is an effort that may well lead to global domination by what Goode describes as an "AI signal" in the future.

The more positive elements of extraterrestrial visitation, e.g. the Nordics, have been covertly assisting the Navy and affiliated elements of the military industrial complex to help build space battle groups with huge vehicles. Tompkins says that the goal is not only to help the Navy establish a security perimeter around our solar system, but to build a broad based alliance with human-looking extraterrestrials concerned about Draconians engaged in imperial space conquest. In addition, these more positive extraterrestrial visitors have been warning secret space program personnel about the dangers posed by Artificial Intelligence.

The similarity of the above two scenarios to key elements in the popular science fiction series *Star Trek* and *Battlestar Galactica* is not accidental. Chapters five and eight presented evidence that Gene Rodenberry (creator of *Star Trek*) and Glenn Larson (creator of *Battlestar Galactica*) were clearly influenced by Leslie Stevens IV (creator of the *Outer Limits*), who was privy to key information about the Navy's secret space program through his father, Admiral Leslie Stevens III. The job of Stevens IV was to sow elements of the truth into the television and motion picture industries as part of a "soft disclosure" program. This dates back to when his father, Admiral Stevens, led a psychological warfare program through the "The Joint Subsidiary Plans Division," which was started by the Pentagon's Joint Chiefs of Staff in 1949. Using the movie/television industry to release some of the truth about secret space programs and extraterrestrial life has been a means of acclimating the public, while concurrently maintaining plausible deniability over what has been disclosed.

Tompkins repeatedly says that he is not a whistleblower since he has been given approval to come forward to reveal his knowledge about the Navy's Solar Warden space program, and the incredibly advanced technologies it uses to deploy its space battle groups. Tompkins says he first received this authority for "full disclosure" in 2001 from Admiral Hugh Webster, who instructed him to "tell it all."[256] To this day, Tompkins continues to receive information from official Navy sources about the Solar Warden space program.

Significantly, three retired Navy officers, Rear Admiral Larry Marsh, Captain Larry Boeck and Commander Art Lumley, have all agreed to publicly identify themselves in stating that Tompkins is a highly regarded and well respected figure in the Navy and its principal support organization, the Navy League. They have all confirmed that Tompkins is an expert in a number of esoteric topics of interest to the Navy. This expertise, according to Boeck and Lumley, extends to Tompkins detailed knowledge and extensive Navy contacts when it comes to the topics of extraterrestrial life and technology.

By Tompkins coming forward to freely reveal his testimony, together with the amount of documentation he has provided, as well as relevant documents released through the National Archives and the Freedom of Information Act, and corroborating statements from retired Navy officers, all this furthers his claim that senior Navy officers approve his "full disclosure" efforts. Consequently, Tompkins testimony and documents go a long way in validating the claims of Corey Goode, who says that "full disclosure" efforts continue to be pursued by a Secret Space Program Alliance that he continues to coordinate with.

This Alliance has been releasing intelligence data about the full range of activities by both multiple secret space programs and extraterrestrial visitors. While Goode and Tompkins' testimonies converge when it comes to full disclosure being approved by the US Navy and other important elements of the global aerospace community, there is also a powerful "limited disclosure" initiative which is supported by the US Air Force.

Endnotes

[222] Interview transcript, "Cosmic Disclosure: SSP Testimonials with William Tompkins," http://spherebeingalliance.com/blog/transcript-cosmic-disclosure-ssp-testimonials-with-william-tompkins.html
[223] William Tompkins, *Selected by Extraterrestrials* (Createspace, 2015).
[224] Goode's first public discussion of the Dark Fleet is available online: Michael Salla, "Secret Space Programs more complex than previously revealed," http://exopolitics.org/secret-space-programs-more-complex-than-previously-revealed/
[225] Cited in "Nazi SS slave empire created through US secret space programs," http://exopolitics.org/nazi-ss-slave-empire-created-through-us-secret-space-programs/
[226] William Tompkins *Selected by Extraterrestrials,* pp. 310-11.
[227] William Tompkins *Selected by Extraterrestrials,* pp. 311.
[228] Tompkins wrote about Admiral Webster's endorsement on the back cover of his book, *Selected by Extraterrestrials.* See http://tinyurl.com/j8sz558
[229] "What is AFCEA?" http://www.afcea.org/site/?q=WhatIsAFCEA
[230] Sue Lannin, "Scientists reverse ageing process in mice; early human trials showing 'promising results,'" http://www.abc.net.au/news/2014-11-04/scientists-reverse-ageing-process-in-mice/5865714
[231] Sue Lannin, "Scientists reverse ageing process in mice; early human trials showing 'promising results,'" http://www.abc.net.au/news/2014-11-04/scientists-reverse-ageing-process-in-mice/5865714
[232] See Michael Salla, "Full ET disclosure plan involves document dumps & whistleblowers coming forward," http://exopolitics.org/full-et-disclosure-plan-involves-document-dumps-whistleblowers-coming-forward/
[233] See "Corey Goode, Intel Update Part 2," http://spherebeingalliance.com/blog/corey-goode-intel-update-part-2-aug-2016.html
[234] Transcript of interview on Gaia TV, September 25, 2016, http://spherebeingalliance.com/blog/transcript-cosmic-disclosure-validating-the-20-and-back-program-with-william-tompkins.html
[235] See Michael Salla, "Recruitment for Covert Service for Secret Space Programs," http://exopolitics.org/recruitment-covert-service-for-secret-space-programs/

---

[236] Transcript of interview on Gaia TV, September 25, 2016, http://spherebeingalliance.com/blog/transcript-cosmic-disclosure-validating-the-20-and-back-program-with-william-tompkins.html

[237] See Michael Salla, "Secret Interstellar Spacecraft use Superluminal Tachyon Drives," http://exopolitics.org/secret-interstellar-spacecraft-use-superluminal-tachyon-drives/

[238] Transcript of interview on Gaia TV, September 25, 2016, http://spherebeingalliance.com/blog/transcript-cosmic-disclosure-validating-the-20-and-back-program-with-william-tompkins.html

[239] Michael Salla, *Insiders Reveal Secret Space Programs and Extraterrestrial Alliances*, pp. 309-44.

[240] Transcript of interview on Gaia TV, September 25, 2016, http://spherebeingalliance.com/blog/transcript-cosmic-disclosure-validating-the-20-and-back-program-with-william-tompkins.html

[241] Michael Salla, "Secret Space Programs more complex than previously revealed," http://exopolitics.org/secret-space-programs-more-complex-than-previously-revealed/

[242] See My interview with Randy Cramer, "Mars Defense Force Defending Human Colonies," http://exopolitics.org/mars-defense-force-defending-human-colonies-interview-transcript-pt-2-2/

[243] See Stephanie Relfe, *The Mars Records,* available online at: http://www.themarsrecords.com/wp/

[244] "Special Access Program Supplement to the National Industrial Security," (Draft 29 May 1992). 3-1-5: *https://fas.org/sgp/library/nispom/sapsup-draft92.pdf* (accessed December 2, 2016).

[245] General Dynamics, "New Construction," http://www.gdeb.com/business_centers/new_construction/

[246] Rense Radio Interview with William Tompkins with Maj. George Filer & Frank Chille – May 4, 2016, http://spherebeingalliance.com/blog/the-amazing-story-continues-part1.html

[247] Corey Goode quoted in "Galactic Human Slave Trade & AI threat to End with Full Disclosure of ET Life," http://exopolitics.org/galactic-human-slave-trade-ai-threat-to-end-with-full-disclosure-of-et-life/

[248] See Corey Goode, "Human Elite Attempt to Negotiate Cessation of SSP Alliance Disclosure in Latest Conference," http://spherebeingalliance.com/blog/human-elite-attempt-to-negotiate-cessation-of-ssp-alliance-disclosure-in-latest-conference.html

[249] Transcript of interview on Gaia TV, October 4, 2016, http://spherebeingalliance.com/blog/transcript-cosmic-disclosure-arrival-of-the-spheres-with-william-tompkins.html

---

[250] Transcript of interview on Gaia TV, October 4, 2016, http://spherebeingalliance.com/blog/transcript-cosmic-disclosure-arrival-of-the-spheres-with-william-tompkins.html

[251] See Michael Salla, "Secret Space War halts as Extraterrestrial Disclosure Plans move forward," http://exopolitics.org/secret-space-war-halts-as-extraterrestrial-disclosure-plans-move-forward/

[252] Transcript of interview on Gaia TV, October 4, 2016, http://spherebeingalliance.com/blog/transcript-cosmic-disclosure-arrival-of-the-spheres-with-william-tompkins.html

[253] Cited in "Extraterrestrial alliance helps secret space program overcome opposition to full disclosure," http://exopolitics.org/extraterrestrial-alliance-helps-secret-space-program-overcome-opposition-to-full-disclosure/

[254] Corey Goode, "Intel Update Part 1," http://spherebeingalliance.com/blog/corey-goode-intel-update-part-1-aug-2016.html

[255] See Michael Salla, "Extraterrestrials Among Us," *Exopolitics Journal*, vol 1:4 (2006). Available online at: http://www.exopoliticsjournal.com/vol-1/1-4-Salla.htm

[256] William Tompkins, *Selected by Extraterrestrials* (Createspace, 2015) backcover.

CHAPTER 10

USAF SANCTIONED LIMITED DISCLOSURE OF ITS SECRET SPACE PROGRAM

Amongst a batch of Wikileaks' email releases from October 2016 were those exposing the identities of two senior military and corporate officials involved in an initiative led by Tom DeLonge, former Blink 182 guitarist and lead singer, to disclose the truth about UFOs and secret space programs. The corporate official was the serving head of Lockheed Martin's Skunkworks, with a long history of building classified aerospace vehicles for the US military intelligence community out of Area 51. The General ran a top Air Force research laboratory out of Wright Patterson Air Force Base up until 2013. This General was shown to be secretly helping DeLonge put together a team of ten senior advisors to tackle the issue of how to reveal to the world the truth about UFOs and a secret space program run by the USAF using antigravity technologies.

The hacked emails are addressed to John Podesta, former Chief of Staff for President Bill Clinton, and Chairman of Hillary Clinton's 2016 Presidential Campaign. Podesta had also earlier been identified by DeLonge as an official working with him to reveal the truth about UFOs. Podesta's involvement in UFO disclosure initiatives is well documented, and spans more than two decades during his service in both the Clinton and Obama Presidential administrations. More recently in December 2015, he got Hillary Clinton to publicly pledge to look into the UFO phenomenon and what was really happening at Area 51. She said: "He [John Podesta] has made me personally pledge we are going to get the information out. One way or another. Maybe we could have, like, a task force to go to Area 51."[257]

Arguably, the most significant among Podesta's many statements[258] about UFOs and government secrecy was his tweet, sent during the final day of his appointment as a Senior Advisor to President Obama on February 13, 2015, where he wrote:

Figure 56. Podesta Tweet on his last day as President Obama's Senior Advisor

John Podesta
@Podesta44
Follow
1. Finally, my biggest failure of 2014: Once again not securing the #disclosure of the UFO files. #thetruthisstilloutthere cc: @NYTimesDowd
8:55 AM - 13 Feb 2015

It is not surprising that Podesta has been secretly corresponding with DeLonge, who gave up his position with Blink-182 to pursue his UFO interest as he explained in an April 2016 *Rolling Stone* interview.[259] What the leaked emails confirm for the first time is the seriousness attached to DeLonge's disclosure efforts by Podesta, and the importance given to it by other very senior people in the US military industrial complex.

The first of the hacked Podesta emails that was released by Wikileaks is dated October 25, 2015 and says:

> Hi John-
>
> Tom DeLonge here, The one who interviewed you for that special documentary not to long ago. Things are moving with the project. The Novels, Films and NonFiction works

> are blooming and finishing. Just had a preliminary meeting with Spielberg's Chief Operating Officer at DreamWorks. More meetings are now on the books-
>
> I would like to bring two very "important" people out to meet you in DC. I think you will find them very interesting, as they were principal leadership relating to our sensitive topic. Both were in charge of most fragile divisions, as it relates to Classified Science and DOD topics. Other words, these are A-Level officials. Worth our time, and as well the investment to bring all the way out to you. I just need 2 hours from you.
>
> Just looking to have a casual, and private conversation in person...
>
> Best,
> Tom DeLonge.[260]

DeLonge is referring to two very important people familiar with "Classified Science and DOD topics," which alludes to advanced technologies related to the UFO issue. DeLonge had earlier met with Podesta in January 2015, to interview him for an upcoming documentary.[261] DeLonge would meet with Podesta again, this time accompanied by two important people. The next meeting actually went ahead according to another leaked email. This email reveals that Podesta was scheduled to meet with DeLonge and a number of "important" people involved in his UFO disclosure initiative the next day, on January 25, 2016.[262]

Another of the hacked emails is dated the day of the January 25 meeting, with the subject header "General McCasland." DeLonge discusses the General:

> He mentioned he's a "skeptic", he's not. I've been working with him for four months. I just got done giving him a four

> hour presentation on the entire project a few weeks ago. Trust me, the advice is already been happening on how to do all this. He just has to say that out loud, but he is very, very aware – as he was in charge of all of the stuff. When Roswell crashed, they shipped it to the laboratory at Wright Patterson Air Force Base. General McCasland was in charge of that exact laboratory up to a couple years ago. He not only knows what I'm trying to achieve, he helped assemble my advisory team. He's a very important man.
>
> Best, Tom DeLonge[263]

In April 2016, a book DeLonge co-authored, *Sekret Machines: Chasing Shadows*, was released. [264] It is the first in a planned multimedia series of books and documentaries promising to blow the lid off of a cover up of secret space programs and the UFO phenomenon. In *Sekret Machines,* DeLonge describes an unnamed General who helped him put together a team of advisors to promote disclosure:

> I've had meetings in mysterious rooms far out in the desert. I've had meetings at the highest levels of NASA. I have had conversations at research centers, think tanks, and even on the phone connected to secret facilities. I've been introduced to a man whom I call "the Scientist," and another whom I call "the General." And there are many more of whom I cannot say much about, but some have become true friends, and all have become close counselors. Each of these men has all held, or currently holds, the highest offices of the military and scientific elite. The point is, I have done it. I have assembled a team of men and women "in the know." And they all believe I am doing something of value, something worth their time and yours.[265]

With the Wikileaks' release, we now know that the unnamed general is Major General William Neil McCasland, who up until 2013 was the Commander of a top Air Force Research Laboratory.[266] What follows is a brief biography of McCasland prior to his retirement:

> Maj. Gen. William N. McCasland is the Commander, Air Force Research Laboratory, Wright-Patterson Air Force Base, Ohio. He is responsible for managing the Air Force's $2.2 billion science and technology program as well as additional customer funded research and development of $2.2 billion. He is also responsible for a global workforce of approximately 10,800 people in the laboratory's component technology directorates, 711th Human Performance Wing and the Air Force Office of Scientific Research.[267]

McCasland's biography clearly establishes that he had the necessary scientific and technical background to be very familiar with the topic of advanced aerospace technologies related to the UFO phenomenon. In earlier chapters, it was pointed out that the foreign and experimental aircraft facilities at Wright-Patterson AFB (then known as Wright Field) was the same location where some of the flying saucer remains from both the 1942 Los Angeles Air Raid and the 1947 Roswell Crash were taken by the Army Air Force.

McCasland, however, is not the only USAF General supporting DeLonge. On the back cover of *Sekret Machines*, an endorsement appears from another retired Air Force General whose last military assignment was Special Assistant to the Commander, Air Force Space Command. Here is what Maj. General Michael Carey had to say:

> Sekret Machines scratches at the surface of "who do" we trust with our classified technology – certainly our adversaries are aware of our undertakings, as they are doing the same, but what of our citizens, our politicians, even our

> own military. Tom DeLonge and A.J. Hartley create a convincing narrative describing the "cat and mouse" game that is timeless between strategic adversaries. It has existed under the sea, on the surface of the earth and in its skies, why wouldn't we believe it occurs in space. Our military leaders have been saying space is a contested environment for years now, perhaps we should believe them! -Maj. Gen. Michael J. Carey.[268]

The endorsement by Carey for Delonge's *Sekret Machines* confirms that at least two senior retired USAF officers are supporting DeLonge's disclosure initiative. Both Generals Carey and McCasland attended the January 25, 2016 meeting with Podesta.

The hacked emails released by Wikileaks also led to the uncovering of the identity of the Lockheed Martin official that attended the January 25 meeting. The hacked January 24 email referred to one of the attendees as "rob.f.weiss@lmco.com" who was tracked down via LinkedIn:

> Rob Weiss can be found on LinkedIn using the email in the WikiLeaks. There, his job title is listed as Executive Vice President & General Manager Advanced Development Programs (Skunk Works) at Lockheed Martin Aeronautics. Skunk Works is the group that built Area 51 at the behest of the CIA, and for decades has worked on developing top-secret advanced aircraft.[269]

Here is how Alejandro Rojas, from *Open Minds TV*, described the significance of the attendees at the January 25 meeting:

> Whoa! Right? DeLonge is getting a meeting with the head of a research lab famous for UFO research, a man who worked for the head of Air Force Space Command, a man who is in charge of the people who run Area 51, and the campaign manager of the person who is, apparently, as of today, likely to be the next president of the United States.[270]

In *Sekret Machines,* Tom DeLonge spends much time describing corporate involvement in developing a secret space program involving the USAF. This book is the aptly packaged product of what the advisory team has been telling Delonge concerning their knowledge of the secret space programs that evolved in the US and Russia. According to DeLonge, their revelations are officially sanctioned by those with "need to know" access. Therefore, this is the most authoritative disclosure yet to emerge on the topics of UFOs, extraterrestrial life and secret space programs.

The hacked Podesta emails released through Wikileaks clearly confirm that DeLonge has very senior insiders from the US military industrial complex helping him move forward in his UFO disclosure initiative. This leads to the all-important question: Is what the advisory team telling DeLonge the full truth about these topics, or only a "limited disclosure" initiative when it comes to secret space programs involving US corporations and different branches of the US military? To get an answer, we need to carefully compare Delonge's depiction of the secret space programs to that which we have learned in previous chapters from the evidence and testimonies presented by William Tompkins and Corey Goode.

The Role of Nazi Germany in Developing a Secret Space Program

In *Sekret Machines,* Delonge divulges information that closely coincides with the "high octane speculations" of the historian, Joseph Farrell, who has closely studied many Nazi-era documents in a series of books, including *The SS Brotherhood of the Bell: Nasa's Nazis, JFK, And Majic-12.*[271] We are told by DeLonge and Farrell that Nazi Germany was partly successful in developing an antigravity "torsion" and/or "scalar" device called the Bell—Die

Glocke. According to Farrell, the Bell was part of the Nazi attempt to develop a super weapon based on scalar physics, whose use in antiquity was described in ancient scriptures.[272]

This is how testing of the Nazi Bell device occurred, as summarized by Nick Cook, author of *The Hunt for Zero Point.* Cook interviewed the original source on the Bell information, who says he saw Nazi documents describing the device:

> The experiments always took place under a thick ceramic cover involved the rapid spinning of two cylinders in opposite directions. The mercury-like substances was code-named "Xerum 525."... The chamber in which the experiments took place was situated in a gallery deep belowground.... its walls were covered with ceramic tiles ... After approximately ten tests, the room was dismantled and its component parts destroyed. Only the Bell itself was preserved.... Each test lasted for approximately one minute. During this period while the Bell emitted its pale blue glow, personnel were kept 150 to 200 meters from it. Electrical equipment anywhere within this radius would usually short-circuit or break down... During the tests, the scientist placed various types of plants, animals and animal tissues in the Bell's sphere of influence. In the initial test period from November to December 1944, almost all the samples were destroyed.[273]

While the Nazi Bell was not successfully weaponized in time to assist the war effort in Europe according to DeLonge, the Bell was secretly transported to South America, and eventually Antarctica in an effort by the Nazis to establish a covert Fourth Reich. The development of the Bell project continued at secret Nazi facilities funded by gold and corporate monies taken out of Germany by Deputy Fuhrer, Martin Bormann, which has been well documented by Paul Manning in the book, *Bormann in Exile,*[274] and more recently by Farrell in *The Third Way: The Nazi International, European Union, and Corporate Fascism.*[275]

In *Sekret Machines,* DeLonge describes how the Nazi Bell project was relocated to Antarctica after South American facilities were compromised by covert allied teams in 1946. The Nazis were able to develop several operational flying saucer craft in Antarctica, which were capable of reaching incredible speeds and were highly maneuverable in the atmosphere.

Operation Highjump, led by Admiral Richard Byrd, was successful in locating the Nazi bases in Antarctica in early 1947 and military hostilities ensued. According to Delonge, although Operation Highjump suffered casualties, it was able to destroy the Nazi facilities and flying saucer prototypes, and take back to the US the recovered booty in terms of scientists, technical information, and resources related to the Nazi Bell research and development program.[276] Here is how DeLonge described one of the dogfights between the Navy's Mustang aircraft and the Nazi UFOs, as narrated by one of the book's characters:

> I had never seen anything airborne move with the agility of the Nazi disk. It rolled and pivoted over the Antarctic mountains, leaping forward and stopping abruptly in mid-air as if gravity had no power over it and aerodynamics were irrelevant. It spun, firing cannon shells, and the Mustangs fell on it like dogs on a bear... For all its speed and maneuverability, the saucer was losing. As it turned to fire on one plane, two more swooped in from behind, riddling it with gunfire, and soon it began to smoke.... Moments later the disk exploded in a fireball that could be seen for miles.[277]

Personnel at the bases of the breakaway Nazis in Antarctica were next described as quickly surrendering. Subsequently, the Antarctica based Nazis no longer provided a significant military challenge to US geopolitical dominance, or to its secret research into antigravity craft using principles of torsion and/or scalar physics.[278]

This is where a March 5, 1947 Chilean press interview with Admiral Byrd, after the premature end of Operation High Jump, raises problems with the scenario DeLonge was told by his advisors. The press report by *El Mercurio* states:

> Adm. Byrd declared today that it was imperative for the United States to initiate immediate defense measures against hostile regions. The Admiral further stated that he didn't want to frighten anyone unduly but it was a bitter reality that in case of a new war the continental United States would be attacked by flying objects which could fly from pole to pole at incredible speeds.[279]

The tone and message that Admiral Byrd delivered in this interview does not appear consistent with someone who has just experienced a decisive military victory against an enemy with flying saucers, but rather, a resounding defeat.

Here, DeLonge's version of events in Antarctica and genesis of a US secret space program diametrically opposes Tompkins and Goode's revelations. According to the latter, Nazi Germany had two secret space programs that ran concurrently during WWII. The Bell experiments and research in Nazi occupied Europe was part of an unsuccessful "wonder weapons" program led by Heinrich Himmler's SS, which was under the direct control of the civil engineer and SS Lieutenant General, Hans Kammler.

By contrast, in Antarctica another secret space program was headed by German Secret Societies that had successfully developed advanced aerospace technologies with the help of two different extraterrestrial races. One was a group of human-looking Nordics, while the other was a species of Reptilians called Draconians. While the Nordics were the first to help the German Secret Societies in the 1920's by providing the designs of antigravity spacecraft through mystics such as Maria Orsic, it was the Draconians in the

1930's who ultimately became more influential with the Nazi regime and German Secret Societies. According to Tompkins, after agreements had been reached with Hitler, the Draconians provided actual working models of flying saucers, and information about underground bases in Antarctica where these models could be reverse engineered during World War II.[280]

Both Tompkins and Goode insist that the German Secret Space Program in Antarctica continued independently, unencumbered by what was happening in Europe under Kammler and the Nazi SS, due to their different goals. The program under Himmler's SS was essentially a weaponization program that was ultimately unsuccessful in using antigravity technologies for the war effort. Consequently, Tompkins and Goode agree with DeLonge and Farrell that Nazi SS efforts to develop weapons technologies such as the Bell were unsuccessful.

At the same time, the German Space Program located in Antarctica had as its main goal to reverse engineer flying saucer designs and prototypes they had acquired for deep space travel. Goode said that the separation between the two German space programs was clearly defined. The Antarctica based program did not share its most advanced weapons research with Himmler's SS, even in the waning days of the war. In response to a question about advanced weapons systems being acquired by the Antarctica based German program to use for the war effort, Goode said:

> I think that there were [some] technologies acquired and integrated into their breakaway secret space program that they were developing. But they were developing this for their own [purposes] when it came down to it, they didn't care about [Germany winning] World War II, the motherland, [or] using this technology to defeat the United States and the enemies they were engaged in war with.[281]

The Germans, in their underground Antarctic bases, subsequently began building fleets of antigravity vehicles capable of deep space missions to the Moon, Mars and beyond. According to Tompkins and Goode, the Reptilians had helped the Germans develop sophisticated weapons platforms for their Antarctica-based spacecraft. This included directed energy weapons that replaced the projectile cannons that the Nazi SS had unsuccessfully attempted to integrate into their own saucer prototypes. The long term goal was that the armed German spacecraft would eventually join Reptilian interstellar fleets in galactic conflicts. Goode has referred to these German space battle groups as the "Dark Fleet," because it remains so highly classified that relatively little is known about it, even to the present day.[282]

Consequently, Operation Highjump was a dismal failure with the German flying saucers comprehensively destroying the Navy's best fighter planes at the time. In an interview, Tompkins revealed that as the naval flotilla approached Queen Maud Land (Neuschwabenland), two waves of flying saucers suddenly came out of the water and engaged the Navy's most advanced fighter planes.[283] The first attack wave comprised 27 flying saucers, and the second 15, which appeared over a 36 hour period according to Tompkins. Some of the saucers had Nazi insignias while others were unmarked. The saucers succeeded in destroying all of the Navy airplanes and helicopters, and sank a destroyer and several other ships. Battered and defeated by the Nazis flying saucer fleets, the rest of the elite Naval aircraft carrier battle group was allowed to turn back for home.

Tompkins and Goode's version of events in Antarctica is supported not only by initial interviews given by Admiral Byrd to the Chilean Press, but also by KGB files citing the ease with which the German flying saucers shot down the Navy aircraft, as reported in two eyewitness accounts. After the Soviet collapse in 1991, the KGB released previously classified files that cast light on Operation Highjump. A 2006 Russian documentary made public for the first time a 1947 secret Soviet intelligence report commissioned by Joseph Stalin on Operation Highjump.[284] In the Soviet intelligence report, never before known testimony was released from two US

Navy servicemen with Operation Highjump. The most telling was by Lieutenant John Sayerson, a flying boat pilot, who is quoted as saying:

> The thing shot vertically out of the water at tremendous velocity, as though pursued by the devil, and flew between the masts [of the ship] at such a high speed that the radio antenna oscillated back and forth in its turbulence. An aircraft [Martin flying-boat] from the Currituck that took off just a few moments later was struck with an unknown type of ray from the object, and almost instantly crashed into the sea near our vessel.... About ten miles away, the torpedo-boat Maddox burst into flames and began to sink... Having personally witnessed this attack by the object that flew out of the sea, all I can say is, it was frightening."[285]

In DeLonge's version of the dogfight between a Nazi flying saucer and the Navy fighters, the flying saucer used cannons. In contrast, according to Sayerson, the Nazi flying saucer used a directed energy weapon that not only took out fighter planes, but also ships.

There is controversy over whether or not a "USS Maddox" served with Operation Highjump, as claimed by Sayerson. Perhaps he was simply mistaken in describing the name of the destroyed torpedo boat, or the name was mistakenly altered in the translation process. Despite unresolved questions over this element in Sayerson's testimony, it is more than likely that the 1947 Soviet report is largely accurate. If so, then the Soviets had learned that the US Navy had suffered a resounding defeat at the hands of the Nazi flying saucer craft hidden in an underground base under the Antarctica ice shelf.

Byrd's Naval expedition returned empty handed, and with a painful lesson that the Antarctica based Germans would be a powerful geo-political force for years to come. Once again, this is a very different account of the Nazi presence in Antarctica to that found in *Sekret Machines*. Why the discrepancy? Before answering,

I need to examine the next major element in *Sekret Machines*—the secret development of the TR-3B "Locust" dual purpose antigravity aircraft and spacecraft.

USAF Space Command's TR-3B Flying Triangles

In *Sekret Machines,* DeLonge spends much time describing the corporate involvement in developing a secret US space program. According to him, a global consortium of corporations secretly funded the development of flying triangles, using antigravity and torsion field principles adopted from the Nazi Bell experiments. These resulted in the development of a squadron of TR-3B's based out of Area 51's highly secretive S-4 facility at Papoose Lake. The US Air Force Space Command and Defense Intelligence Agency is in charge of the TR-3B or 'Locust', operating out of S-4, according to Delonge via his advisory team.

Delonge went on to explain in *Sekret Machines* that the TR-3B is a combined US military and corporate made aerospace vehicle capable of operating both near the Earth's surface as a conventional aircraft, and in near Earth orbit as a spacecraft. He described different sized TR-3B vehicles, with the largest being several hundred feet across.

DeLonge's information about the TR-3B closely matches another earlier account by Edgar Fouche, an aerospace engineer, who worked at Area 51. While there, he learned about the TR-3B's existence as the most highly classified aerospace vehicle built by the US military industrial complex in the late 1980's. In 1998, Fouche first came forward to reveal the TR-3B's existence; that it was stationed at the S-4 facility, and was as much as 600 feet wide, which is similar to what Delonge later claims he was told by his advisory team.[286]

Documents provided by Fouche add much credibility to his testimony, which is also supported by the many flying triangle sightings reported in Belgium, and elsewhere.[287] Indeed, *Sekret Machines* explains the 1989 Belgium sightings as directly related to flights of the TR-3B.[288]

Figure 57. Illustration of TR-3B. Source: Edgar Fouche

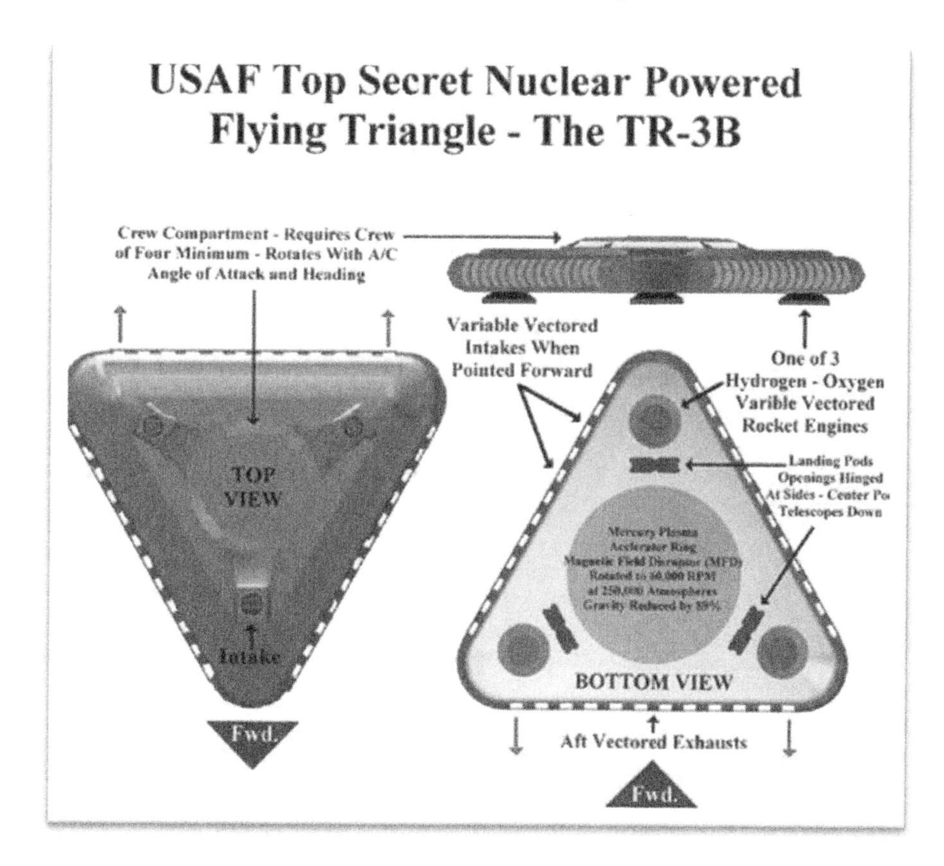

Once again, the testimony of Tompkins and Goode differs dramatically to DeLonge's information, as provided to him by his advisory team. In chapter four, I presented Tompkins' information concerning his direct involvement in the design of space battle groups that would be under the control of the Navy. To briefly summarize, in the 1950's Tompkins conceptualized different designs for kilometers-long space carriers and cruisers. The space battle groups were constructed in massive underground facilities in the Wasatch Mountains of Utah, and were first deployed in 1984 as the Solar Warden program. Eight battle groups were eventually constructed, and these rotate with four in service at any time, while the other four undergo maintenance—similar to the scheduling of aircraft carriers.

Goode has said that he directly served in Solar Warden from 1987 to 2007. Significantly, he says that the TR-3B became outdated with the deployment of Solar Warden space fleets, and was allotted as a hand-me-down to less classified programs run by the Air Force and other groups:

> There were more than 8 of the "Cigar" shaped carrier class and other class of vessels of various sizes and classifications. They were designed to carry various type of "craft" that many think of as the TR-3B... The TR-3B is considered extremely outdated technology and in many cases has been gifted as "hand me down" craft to "Elites" within the Secret Earth Governments and their Syndicates as something akin to "Company Jets". There are so many newer technologies that are of the same general shape as the TR-3B (& models that came after) that it would blow people's minds.[289]

Consequently, what Goode and Tompkins' testimonies show is that the TR-3B is part of a second tier US secret space program run out of Area 51.

Are Extraterrestrials Real?

In *Sekret Machines,* an incident is described where an apparent UFO abduction occurs, but the supposed "Grey" extraterrestrial is unmasked by its female victim and shown to be a human disguised in an alien costume.[290] After escaping the remote facility where she is being held, the victim learns that she had been taken to Siberia by a covert Russian group.[291] The abduction incident is depicted to be part of a Russian secret space program, which uses technologically advanced arrow-shaped aerospace vehicles similar in flight performance to the TR-3B.[292]

Delonge describes other UFO incidents involving covert Russian spacecraft. These include UFO incidents at US nuclear facilities which led to them being deactivated. The Russian spacecraft were allegedly interfering with the US nukes as a means of demonstrating Russia's ability to deactivate and even destroy America's nuclear arsenal. Consequently, it is not extraterrestrials behind many UFO incidents such as abductions and nuclear weapons deactivation, but Russian secret spacecraft according to DeLonge's version, which is based on information from his advisory team. Especially relevant here is the language General Carey used in his endorsement of *Sekret Machines*:

> Tom DeLonge and A.J. Hartley create a convincing narrative describing the "cat and mouse" game that is timeless between strategic adversaries. It has existed under the sea, on the surface of the earth and in its skies, why wouldn't we believe it occurs in space.[293]

He is all but naming Russia as the strategic adversary that has long battled the US in all strategic theaters of operation, including more recently, space.

Once again, there is a dramatic difference here with what Tompkins and Goode have to say about extraterrestrials and secret space programs, especially when it comes to long term strategic threats. Earlier, I commented on their accounts of Nazi Germany being assisted by Reptilian and Nordic extraterrestrial groups at different stages in their development of two secret space programs. Later, according to Tompkins, Nordic extraterrestrials assisted the US Navy program to establish space battle groups. The Nordics were seeking allies in their galactic conflict with the Draconian

Reptilians, and believed that the Navy would eventually become one.

Tompkins says that Nordic extraterrestrials were embedded in aerospace companies where he worked, such as Douglas Aircraft Company and TRW, and actively helped in the research and development of antigravity spacecraft. He asserts that the Nordics helped counter the industrial sabotage by Reptilians in both the NASA and the nascent Navy secret space program.

Goode also says that the Solar Warden and Dark Fleet programs coordinate with different extraterrestrial visitors. In addition, he states that the corporate run space program has industrial scale manufacturing centers on Mars that trade with up to 900 extraterrestrial civilizations.[294] DeLonge, however, proposes a wildcard to explain stories of encounters with advanced extraterrestrial visitors and the discoveries of non-human constructed spacecraft. He refers to mythological accounts of the "Gods" as referred to throughout history, and suggests these were not flights of fancy by ancient peoples, but real beings that manipulated humanity for both positive and negative agendas.[295]

In Homer's *Illiad,* for example, the reader of this epic poem learns how the Gods took different sides in this war in ancient times between the Greeks and Trojans—the superpowers of that region of the world back then. The conflict was wholly contrived by a mischievous Goddess called Eris, who rolled her "golden apple of discord" at the foot of the Gods, and set off a chain of events that led to the Trojan War.

Sekret Machines ends on a cliffhanger regarding the role of the Gods and current secret space programs. It is expected that future volumes will describe different ways in which the Gods have helped foment geo-political conflict between major world powers such as the US and Russia—the modern equivalents to Ancient Greece and Troy. This is likely to involve the recovery of highly advanced non-human spacecraft at various crash retrieval sites, which DeLonge attributes to the Gods. These are modern day versions of the golden apples of discord that are planted by the Gods.

Limited versus Full Disclosure

How does one account for the significant differences between the secret space program disclosures by Delonge and his advisory team, and those of Tompkins and Goode? If one accepts Tompkins and Goode's testimonies as fundamentally accurate, as my analysis of the documentary evidence in this book suggests, then there are two scenarios which come to mind to explain these discrepancies.

The first is that we are dealing with a "limited hangout," which according to Victor Marchetti, a former Special Assistant to the CIA Deputy Director, is "spy jargon for a favorite and frequently used gimmick of the clandestine professionals." [296] Marchetti went on to describe how a limited hangout works:

> When their veil of secrecy is shredded and they can no longer rely on a phony cover story to misinform the public, they resort to admitting—sometimes even volunteering—some of the truth while still managing to withhold the key and damaging facts in the case. The public, however, is usually so intrigued by the new information that it never thinks to pursue the matter further. [297]

A limited hangout scenario suggests that DeLonge is being fed a well-crafted narrative by his advisory team and the intelligence community, who believe their cover stories for the UFO phenomenon no longer have credibility.[298] The new narrative is designed to appeal to widespread public belief in a UFO cover-up, by revealing some of the advanced technologies developed in deep black programs, which would account for the UFO phenomenon dating back to the World War II era.

While the public learns of antigravity spacecraft, they are told that the extraterrestrial hypothesis—that UFOs are interplanetary spacecraft—is false. The only 'alien' factor involved here are the elusive Gods mentioned in historical texts who today are manipulating the US and Russia into a nuclear war, or planting advanced spacecraft in remote locations as "golden apples of discord." The preceding scenario will appeal not only to scientific skeptics, but also to followers from different religions who can view this information as confirming the legitimacy of their scriptures.

What this scenario conveys is that DeLonge is being given access to genuine information that is part of a limited hangout operation by senior members of the US military industrial complex. We know from the leaked emails released by Wikileaks that his advisory team includes retired USAF Major Generals McCasland and Carey, and the current head of Lockeheed's Skunkworks, Rob Weiss. These three, plus the seven others making up DeLonge's advisory team, appear to be carefully crafting a narrative through DeLonge that maintains the legitimacy of the military industrial complex. It is modeled for some kind of official acknowledgement of a secret space program based at Area 51 which uses the TR-3B.

After all, if there are Gods hidden in the shadows manipulating both Russia and the US into having a nuclear conflict over advanced technologies, then the whole issue concerning secret space programs becomes firmly embedded in national security concerns. It is important to keep in mind that this narrative was not only being fed to DeLonge, but also through him to Podesta, and then certainly on to the unsuccessful 2016 Presidential candidate, Hillary Clinton, who also has a long association with the UFO topic dating back to Bill Clinton's Presidential administration.[299]

A second scenario to be explored is that Delonge's advisors, who we know are very senior figures in the military industrial complex, are simply out of the loop when it comes to the most advanced secret space program technologies. Delonge has repeatedly said that his advisors work at or are familiar with Area 51 facilities, where the TR-3B is based. We know that is certainly the case for Levin, since Skunkworks has had a long historical relationship with Area 51, dating back to its founding in the 1950's. Also, Major General McCasland would have certainly been aware of Area 51 projects, given the long historical relationship between Area 51 and the research laboratories at Wright Patterson Air Force Base when it comes to studying retrieved UFOs.

According to Tompkins, however, it is in large underground facilities in Utah's Wasatch Mountains where the Navy's space battle fleets have been secretly constructed. Tompkins says that the construction facilities employ the same modular construction technique used for the building of the navy's aircraft carriers and submarines at Newport News Shipbuilding facilities, which from 2001 to 2009, was owned by Northrup Grumman. He explains that Northrup Grumman is the primary corporation for building the Navy's secret space battle craft using similar construction techniques to those used to build aircraft carriers and nuclear submarines.

What this material implies is that there is a secret space program run by military entities such as US Air Force Space Command, the Defense Intelligence Agency (DIA), and the National Reconnaissance Office (NRO), which is entirely separate to the program the US Navy has been doing in the Wasatch Mountains with the assistance of select corporate contractors. The compartmentalization process is so successful between the different space programs that USAF/DIA/NRO personnel do not believe that such a Navy run space program exists.

Such views date back to the creation of Project RAND in October 1945. As presented in chapter three, Army Air Force leaders such as General Curtis LeMay "considered space operations to be an extension of air operations."[300] This led to the Navy being pushed into a secondary support role in developing future generations of satellites, which came under the jurisdiction of the

USAF, and the National Reconnaissance Office when it was created in 1961. This meant that dealing with UFOs and the extraterrestrial phenomenon would come under the purview of the USAF and the NRO. Put simply, the USAF and NRO would dominate space affairs, while the Navy would remain focused on what was happening on and beneath the world's oceans.

The RAND Corporation subsequently worked closely with the USAF in developing and implementing a coordinated space policy dealing with the extraterrestrial phenomenon, with the Navy apparently doing little more than just monitoring the situation. As William Tompkins attests, however, little did USAF leaders realize that the Navy had not conceded space affairs to the Air Force at all!

Instead, while the USAF and NRO worked on building satellites, spy planes, stealth space stations, and squadrons of antigravity craft out of Area 51 for near Earth space operations, the Navy worked on large battle groups for deep space operations out of its Wasatch mountains facilities. While the USAF and NRO conducted research and development out of state of the art aerospace laboratories at Wright Patterson Air Force Base, the Navy had its own, arguably more advanced, laboratories at China Lake. It is important to remember that Tompkins said he travelled to China Lake approximately 40 times during the war, carrying briefing packets about the Nazi space programs and different extraterrestrial visitors.

Consequently, it is not surprising that personnel participating in a secret space program run by the USAF and NRO, along with the DIA, would not know about the Navy's space program. This is vividly illustrated in abduction incidents described by Goode. He says that sometime in January-February 2016, he was abducted at least three times by covert teams associated with the USAF/NRO/DIA who interrogated him.[301] This was followed with him being abducted again by the same group in late September 2016, where a more senior USAF officer interrogated him.[302]

Apparently, some of the intelligence released since September 2014 by Goode, concerning secret space programs, was accurate enough to come to the attention of a "lower level secret space program" (Goode's term for it) run by the USAF Space Command and its NRO and DIA partners. Subsequently, he was subjected to a military abduction by individuals he concluded were part of the USAF Space Command.[303] He described the September 2016 incident involving the more senior USAF officer:

> As I was focused on what they were doing, another person came into the room. The two airmen snapped to attention. All three of the men were wearing Air Force uniforms without any patches to indicate who they were. The last person to enter the room had white hair, a white goatee beard, and a very serious look on his face. He sat in the chair next to me while the two airmen once again took hair and blood samples from me. After putting away the samples, they were told to leave the room. The man who was obviously in charge addressed me by my first name. He began to ask random questions about the LOC [Lunar Operations Command] and the status of the SSP Alliance.[304]

Goode stated that he experienced the military abductions on a spacecraft that was clearly less advanced than those belonging to what he has described as a Secret Space Program Alliance, which includes significant elements of the Navy's Solar Warden program.[305]

Goode's interrogators could not find any information about him in their databases, which is not surprising given the level of compartmentalization involved in different unacknowledged Special Access Programs. Cover stories and other methods are used to hide these programs from those without "need to know" access.[306] In this regard, a 1992 supplement to a Department of Defense document titled: "National Industrial Security Program Operating Manual," states the need for cover stories when it comes to unacknowledged programs:

> Program Cover stories. (UNACKNOWLEDGED Program). Cover stories may be established for unacknowledged programs in order to protect the integrity of the program from individuals who do not have a need to know. Cover stories must be believable and cannot reveal any information regarding the true nature of the contract. Cover stories for Special Access Programs must have the approval of PSO [Program Security Officer] prior to dissemination.[307]

Figure 58. USAF/DIA/NRO Spacecraft which landed near Corey Goode's home. Credit: Gaia TV

Goode said that all his records were scrubbed upon returning to civilian life in 1986, after he was age-regressed and sent back in time after completing his 20 year tour of duty in 2007.[308] During one of Goode's interrogations, attempts were made to brainwash him into believing that extraterrestrials weren't real and that the "higher level space programs" he was discussing—Solar Warden, Interplanetary Corporate Conglomerate and Dark Fleet—did not exist.[309] In an interview, Goode recalled what one of his Air Force interrogators said:

> One of them began telling me, "You know that all of these beings are not really aliens. They're just us from the future from two different timelines." And I couldn't really respond. And he was saying that the Nordic type of people are us from the future. And these little grey beings are us from a different timeline in the future. And they are coming back to try to fight over the timelines. And they were stating that the Greys, their genetics had broken down so badly in the future that they were coming back to get genetic stock as well.[310]

According to this interpretation of the extraterrestrial phenomenon, all alleged contact cases involve humans from the future, either Nordics or Greys, which have genetically evolved from present day humanity. All this is very significant given what DeLonge has been told by his advisory team, and what presumably he passed on to John Podesta about the secret space program out of Area 51.

Essentially, the military services involved in Goode's abductions are convinced that his disclosures about extraterrestrial life are part of an elaborate disinformation campaign by an Unacknowledged Special Access Program. As mentioned earlier, personnel in these programs are instructed to use cover stories to hide the truth of their activities. Goode's abductors clearly believed that his secret space program disclosures were an elaborate cover story. It's noteworthy that the USAF officer that interrogated him during the September 2016 abduction was highly skeptical of his story, as Goode narrated:

> I asked how I should address him and he looked at me as if I was an idiot. He said, "Just call me Sir." I asked him why I was being picked up again, and why he was so skeptical after the last time when I was tricked into outing three members of the SSP Alliance. He stated he was here to repeat the process of gathering evidence and then keep the

chain of custody of the evidence. He stated this was to rule out the possibility that the previous team had made a mistake or had tampered with the results.

He went on to say that he could not accept that I was telling the truth. The information I had provided was too far outside the scope of his briefings. He stated that it was possible that the SSP Alliance was merely a few people from his own program who had gone rogue.[311]

Goode's recollections of the USAF officer's statements indicate that full disclosure is not just a matter of the general public learning about the broader truth behind secret space programs and extraterrestrial visitors, but also for military and corporate personnel to come up to speed. The compartmentalization process means that what is considered "full disclosure" by one level of the secrecy system is but a "limited disclosure" at more highly classified levels.

Figure 59. Edward Snowden leaked NSA slide on Sentry Eagle

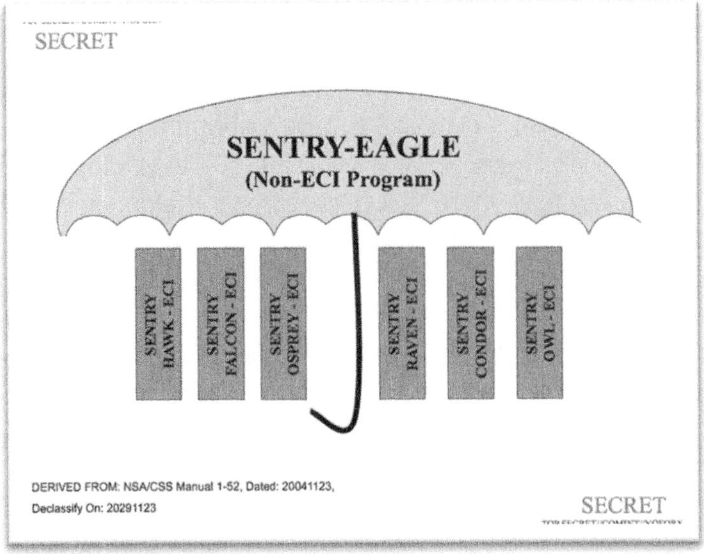

This is illustrated by the above NSA leaked slide showing how different DHS, DOD, and NSA classified programs are hidden within less classified programs. Due to compartmentalization, those in less classified 'cover' programs would be out of the loop on the more highly classified programs. Consequently, the scenario described by Goode in terms of his abduction by USAF Space Command personnel is entirely plausible.

Conclusion

Both Goode and Tompkin's testimonies portray a deeper layer of secret space program operations and technologies, far exceeding anything happening in near Earth orbit under the control of the USAF/DIA/NRO. Due to compartmentalization, this deeper level space program, Solar Warden, is not known by personnel working on antigravity spacecraft such as TR-3B under the USAF/DIA/NRO. It is therefore very feasible that corporations and personnel working on classified aerospace programs at Area 51 are out of the loop on the Solar Warden program.

It is also plausible that the advisory team of ten senior corporate and military personnel feeding DeLonge intelligence may be part of a limited hangout operation. They are intentionally spilling the beans on what has been taking place at Area 51, but only in order to keep hidden even more important secret programs happening elsewhere. Alternatively, DeLonge's advisory team may be genuinely in the dark about the Navy led Secret Space Program based out of the Wasatch Mountains, and its current operations in deep space.

Despite the question of which of the above two scenarios better explains what is happening with DeLonge and his advisors, the disclosures in *Sekret Machines* genuinely reveal the first layer of a multilayered set of secret space programs. Wikileaks documents prove that covert leaders in the USAF Secret Space Program have chosen DeLonge to be part of an officially sanctioned disclosure initiative. DeLonge's initiative became a factor in the 2016 US Presidential election through Podesta's involvement, and his chairmanship of the Clinton Presidential Campaign. Unfortunately, for Podesta and Clinton, Donald Trump surprised everyone with his election win.

Endnotes

---

[257] Daymond Steer, "Clinton promises to investigate UFOs," http://www.conwaydailysun.com/newsx/local-news/123978-clinton-promises-to-investigate-ufos
[258] For earlier statements by Podesta on the UFO issue, see "Obama White House denied access to UFO files," http://exopolitics.org/obama-white-house-denied-access-to-ufo-files/
[259] Patrick Doyle, "Inside Tom DeLonge's UFO Obsession, Blink-182 Turmoil," *Rolling Stone*. http://www.rollingstone.com/music/news/inside-tom-delonges-ufo-obsession-blink-182-turmoil-20160427
[260] "Important Things," https://wikileaks.org/podesta-emails/emailid/2125
[261] See Alejandro Rojas, "Hillary campaign manager held UFO meeting with USAF generals, rock star and top secret aircraft developer," http://www.openminds.tv/hillary-campaign-manager-held-ufo-meeting-with-usaf-generals-rock-star-and-top-secret-aircraft-developer/38698
[262] The email refers to an attempt by one of the attendees to confirm the exact meeting time, "Re: Invitation: DeLonge/Podesta Meeting @ Mon Jan 25, 2016 10:30am - 11:30am" (neilmcc79@gmail.com) https://wikileaks.org/podesta-emails/emailid/5078
[263] The leaked email is titled "General McCasland," https://wikileaks.org/podesta-emails/emailid/3099
[264] Tom DeLonge and A.J. Hartley, *Sekret Machines Book 1: Chasing Shadows* (To The Stars, 2016)
[265] Tom DeLonge and A.J. Hartley, *Sekret Machines*, Kindle Locations 95-100
[266] The leaked email is titled "General McCasland," https://wikileaks.org/podesta-emails/emailid/3099
[267] US Air Force, Major General William N. McCasland, http://www.af.mil/AboutUs/Biographies/Display/tabid/225/Article/104776/major-general-william-n-mccasland.aspx
[268] Tom DeLonge and A.J. Hartley, *Sekret Machines,* backcover.
[269] Alejandro Rojas, "Hillary campaign manager held UFO meeting with USAF generals, rock star and top secret aircraft developer," http://www.openminds.tv/hillary-campaign-manager-held-ufo-meeting-with-usaf-generals-rock-star-and-top-secret-aircraft-developer/38698
[270] Alejandro Rojas, "Hillary campaign manager held UFO meeting with USAF generals, rock star and top secret aircraft developer,"

---

http://www.openminds.tv/hillary-campaign-manager-held-ufo-meeting-with-usaf-generals-rock-star-and-top-secret-aircraft-developer/38698

[271] See Joseph Farrell, *The SS Brotherhood of the Bell: Nasa's Nazis, JFK, And Majic-12* (Adventures Unlimited Press, 2006); and *Nazi International: The Nazis' Postwar Plan to Control the Worlds of Science, Finance, Space, and Conflict* (Adventures Unlimited Press, 2009).

[272] See Joseph Farrell, The Cosmic War: Interplanetary Warfare, Modern Physics and Ancient Texts (Adventures Unlimited Press, 2007).

[273] Nick Cook, *The Hunt for Zero Point: Inside the Classified World of Antigravity Technology* (Broadway Books, 2001) 192.

[274] Paul Manning, *Martin Bormann: Nazi in exile* (Createspace, 1981).

[275] Joseph Farrell, *The Third Way: The Nazi International, European Union, and Corporate Fascism* (Adventures Unlimited Press, 2015)

[276] For DeLonge's account of the battle in Artarctica and the destruction of the Nazi facilities, see *Sekret Machines*, ch. 52.

[277] Tom DeLonge & A.J. Hartley, *Sekret Machines* (Kindle Location 6501).

[278] For discussion of Scalar Physics, see Joseph Farrell, The Cosmic War: Interplanetary Warfare, Modern Physics and Ancient Texts, 28-66. For video presentation by Richard Hoagland discussing Torsion Field physics, go to: https://archive.org/details/TorsionPhysics2013

[279] Quoted in an interview of Admiral Byrd by Lee van Atta, "On Board the Mount Olympus on the High Seas" *El Mercurio*, (Santiago, Chile, March 5, 1947). See "The Antarctic Enigma," http://www.bibliotecapleyades.net/tierra_hueca/esp_tierra_hueca_6c.htm

[280] Chapter two discusses Tompkins account of the agreements reached between Hitler and the Reptilians. For transcript of interview where Tompkins discusses the alliance, go to: http://exopolitics.org/interview-transcript-us-navy-spies-learned-of-nazi-alliance-with-reptilian-extraterrestrials/

[281] See transcript of Corey Goode interview, "Cosmic Disclosure: Raiders of Lost Technology, Season 2, Episode 5," http://spherebeingalliance.com/blog/transcript-cosmic-disclosure-lost-technology.html

[282] Corey Goode interview in "Cosmic Disclosure: The Dark Fleet, Season 4, Episode 6," http://spherebeingalliance.com/blog/transcript-cosmic-disclosure-the-dark-fleet.html

[283] Private interview with William Tompkins, date/details?

[284] See "Third Reich - Operation UFO (Nazi Base In Antarctica) Complete Documentary," https://youtu.be/MwUpPwyyvLw

---

[285] Frank Joseph, Our Real "War of the Worlds," New Dawn Special Issue Vol.6 No.5. Available online at: http://www.newdawnmagazine.com/special-issues/new-dawn-special-issue-vol-6-no-5
[286] See Edgar Fouche, "Presentation to IUFOR by Edgar Rothshild Fouche." http://tinyurl.com/z665g7y
[287] See Edgar Fouche, "Presentation to IUFOR by Edgar Rothshild Fouche." http://tinyurl.com/z665g7y
[288] For DeLonge's account of TR-3B's in Belgium, see *Sekret Machines*, ch. 28.
[289] See the following interview with Corey Goode, "Secret space programs more complex than previously revealed," http://exopolitics.org/secret-space-programs-more-complex-than-previously-revealed/
[290] For DeLonge's account of the staged alien abduction in Russia, see *Sekret Machines*, chs. 41 & 45.
[291] See *Sekret Machines*, ch 45.
[292] For DeLonge's account of a fictional dog fight between a TR-3B and its Russian equivalent, see *Sekret Machines*, ch. 39.
[293] Tom DeLonge and A.J. Hartley, *Sekret Machines,* backcover.
[294] See Corey Goode, "Joint SSP, Sphere Alliance & ICC Leadership Conference & Tour of Mars Colony on 6.20.2015: ICC Mars Colony Conference Report June 20th 2015," http://spherebeingalliance.com/blog/joint-ssp-sphere-alliance-icc-leadership-conference-tour-of-mars-colony-on-6-20.html
[295] For DeLonge's account of Greek Legend and the Gods, see *Sekret Machines*, ch. 83.
[296] Victor Marchetti (August 14, 1978) The Spotlight. Cited online at https://en.wikipedia.org/wiki/Limited_hangout
[297] Victor Marchetti (August 14, 1978) The Spotlight. Cited online at https://en.wikipedia.org/wiki/Limited_hangout
[298] Another critique of DeLonge that examines this further is Robbie Graham, "The DeLonge DeLusion: Part One," http://mysteriousuniverse.org/2016/11/the-delonge-delusion-part-one/
[299] See articles by Grant Cameron, "The Presidents UFO Website," http://www.presidentialufo.com/bill-clinton
[300] "Preliminary Design of an Experimental World-Circling Spaceship," http://www.rand.org/pubs/special_memoranda/SM11827.html
[301] Goode gave me a private briefing on his abductions which was summarized in the following article: "Military Abduction & Extraterrestrial Contact Treaty – Corey Goode Briefing Pt 2," http://exopolitics.org/military-abduction-extraterrestrial-contact-treaty-corey-goode-briefing-pt-2/ See also Corey Goode, "Latest Intel and Update," http://spherebeingalliance.com/blog/latest-intel-and-update.html
[302] Corey Goode interview, "Cosmic Disclosure: Troubling Encounters,"

---

http://spherebeingalliance.com/blog/transcript-cosmic-disclosure-troubling-encounters.html

[303] Goode explained why he believed his abductors in the early 2016 incidents were USAF in an interview, see "Cosmic Disclosure: Veiled Threats and Open Disclosures, Season 5, Episode 7," http://spherebeingalliance.com/blog/transcript-cosmic-disclosure-veiled-threats-and-open-disclosures.html

[304] Corey Goode, "Are We Navigating To Our Optimal Temporal Reality?" http://spherebeingalliance.com/blog/are-we-navigating-to-our-optimal-temporal-reality.html

[305] Goode's testimony about different space programs was studied in detailed in my book, *Insiders Reveal Secret Space Programs and Extraterrestrial Alliances* (2015). For a short online article discussing these, go to; "Secret space programs more complex than previously revealed," http://exopolitics.org/secret-space-programs-more-complex-than-previously-revealed/

[306] See Michael Salla, "Astropolitics and the "Exopolitics" of Unacknowledged Activities in Outer Space," *Astropolitics: The International Journal of Space Politics & Policy*, Vol. 12:1 (2014). An earlier version is available online at: http://exopolitics.org/astropolitics-and-the-exopolitics-of-unacknowledged-space-activities/

[307] "Special Access Program Supplement to the National Industrial Security," (Draft 29 May 1992). 3-1-5: *https://fas.org/sgp/library/nispom/sapsup-draft92.pdf* (accessed December 2, 2016).

[308] For a summary of Goode's recruitment and service in a 20 year and back program, see "Recruitment & Covert Service for Secret Space Programs," http://exopolitics.org/recruitment-covert-service-for-secret-space-programs/

[309] For a short online article discussing these, go to; "Secret space programs more complex than previously revealed," http://exopolitics.org/secret-space-programs-more-complex-than-previously-revealed/

[310] Corey Goode, "Cosmic Disclosure: Veiled Threats and Open Disclosures, Season 5, Episode 7," http://spherebeingalliance.com/blog/transcript-cosmic-disclosure-veiled-threats-and-open-disclosures.html

[311] Corey Goode, "Are We Navigating To Our Optimal Temporal Reality?" http://spherebeingalliance.com/blog/are-we-navigating-to-our-optimal-temporal-reality.html

CHAPTER 11

PILLARS OF DISCLOSURE: PRESIDENT TRUMP'S RELATIONSHIPS WITH THE NAVY, FBI & RUSSIA

On November 8, 2016, Donald Trump surprised most political pundits and media networks with his Presidential election victory over Hillary Clinton. Trump struck a chord with middle class Americans, who had become disenchanted with the globalization process and the resulting loss of high-paying manufacturing jobs. Trump's plan for reinvigorating the country's manufacturing industry led to widespread support from millions of Americans fearful of what the future would bring if the political status quo was maintained. His victory requires an examination of how his administration will impact the prospects for "official disclosure" of secret space programs and extraterrestrial life.

There are three key institutional relationships that are very important for anticipating the policies of a Trump Presidential administration, especially when it comes to some kind of official disclosure. First, during the Presidential campaign, Trump said that he would rebuild the Navy, whose ocean-going fleets had been radically impacted by budgetary cutbacks. Second, Trump was directly helped by the Director of the Federal Bureau of Investigations (FBI), James Comey, whose brief reopening of an investigation of Hillary Clinton only two weeks before the Presidential elections played a decisive role in his victory.[312] Finally, Trump's positive references to Russia and its leader, President Vladimir Putin, indicated that he wants to radically change the present strained relations between the US and Russia in a manner that would significantly impact global affairs.

The ways in which Trump develops his administration's institutional relationships with the Navy, the FBI and the Russian Federation will be very significant in terms of how much he will be able to publicly disclose about the secret space programs and extraterrestrial life. If these relationships are properly cultivated, then his administration may go well beyond DeLonge's "limited disclosure" concerning the USAF's secret space program, and create the conditions for "full disclosure" of all secret space programs to occur. Full disclosure, in particular, would not only dramatically reinvigorate the US manufacturing industry in building advanced technologies for mass public consumption, but lead to revolutionary global changes.

A Manhattan Project to Rebuild the Navy's [Space] Fleets

One of the core plans unveiled by Trump during the Presidential campaign was to reinvigorate the Navy and its ship building industry. On October 21, 2016, Trump made a major policy speech in Pennsylvania outlying his grand plan:

> Our Navy is the smallest it's been since World War I. My plan will build the 350 ship Navy we need. This will be the largest effort at rebuilding our military since Ronald Reagan, and it will require a truly national effort. The Philadelphia Navy Yard is a perfect example. I will instruct my Secretary of the Navy to study locations like Philadelphia with a long

> history of service to our military and proximity to vibrant private industry and find ways to involve them in this national effort.
>
> As our fleet is rebuilt, we'll need to invest in recruiting the skilled American craftsmen we need, like welders and pipe fitters and so much more. We will establish 'centers of excellence' in places like Philadelphia and Portsmouth, New Hampshire and Hampton Roads in Virginia to produce the master craftsmen we need to rebuild our Fleet. We will rebuild our navy and we will do it with American steel made right here in Pennsylvania.[313]

In a memo, Trump's senior military defense advisor, Alexander Gray, referred to the plan to rebuild the Navy's fleets as a new "Manhattan Project":

> Donald Trump has promised to rebuild America with American hands and American steel. On Day One of his administration, he will immediately begin to fulfill that promise with a Manhattan Project urgency to rebuild our navy, which has shrunk under the Obama-Clinton years to its lowest level since World War I.[314]

Trump's Navy policy has proved to be popular, judging by its support by retired Navy officers. Among the list of 88 flag officers who signed a letter supporting Trump's candidacy, the most signatures came from the Navy; 34 Admirals signed the petition, along with an additional two Coast Guard Admirals, and two Marine Corps Generals.[315]

Trump's Navy policy promises to be truly groundbreaking in terms of what it will open up for the Navy's future development of its Solar Warden Secret Space Program. The Navy's eight battle groups, secretly built and deployed in the 1980's, are currently operational. However, according to William Tompkins, the Navy plans to replace these existing battle groups, which are becoming antiquated in comparison with other space programs, with twelve new battle groups that are currently under development. These are tentatively scheduled to become fully operational in the 2030's, replicating the 50 year lifecycle of modern aircraft carriers. In a February 25, 2016 interview exchange between Tompkins [WT] and myself [MS], he elaborated upon the Navy's plans.

> [WT] All of the main eight Solar Warden battle groups are old, real old. They are in the process of being completely replaced by 12 new ones. Which is a major program going on, because they are very old.
>
> [MS] So the current plan involves 12 battle groups under ... development.
>
> [WT] They are completely new, everything is new. Not just the ship ... [new] hardware, everything.
>
> [MS] When will they be deployed?
>
> [WT] – I think first deployment will be two [battle groups] in 2031, or close.
>
> [MS] – They are under current design and manufacture ...
>
> [WT] – It's all part of the stage. They have built a real fast program to get a prototype. They are four kilometers long rather that one kilometer.[316]

Such an undertaking will clearly require enormous funds and resources that would strain the Navy's share of the Pentagon's "deep black budget," which was estimated to be $1.7 trillion per year by the end of the Clinton administration in January 2001.[317] To fully appreciate the significance of such a vast sum, consider that the proposed Pentagon budget for 2017 was only $582 billion.[318] This means that the deep black budget used by the Navy, Air Force and other secret space programs was, back in the year 2000, *three times the current Pentagon budget today*! Trump's support for rebuilding the Navy surface fleets with "Manhattan Project urgency" comes as a clear signal by him and his senior advisors that they would be receptive to a policy of prioritizing the construction of a new generation of kilometers-long spacecraft for the Navy's Solar Warden program.

This raises the question of whether Trump was covertly supported by groups allied with the Solar Warden program and other connected government agencies during the Presidential Campaign. The Wikileaks' documents, for example, greatly assisted the Trump campaign by exposing much about the shortcomings of Hillary Clinton and her senior campaign advisors. While Clinton accused Russia of being behind the leaks,[319] the real culprit, according to Corey Goode, was an "Earth Alliance" that coordinates with the Navy's Solar Warden program.[320]

Goode says that the Earth Alliance comprises multiple factions and organizations involved in various supply, support and command roles when it comes to the different secret space programs. The Earth Alliance includes "White Hats" among the military industrial elite, the intelligence community of the US, and many other major countries that are supportive of the goals of the Solar Warden program, along with its allies that make up what Goode describes as the "Secret Space Program Alliance." In addition, the Earth Alliance includes the BRICS nations (Brazil, Russia, India, China, South Africa) which are intent on creating a new global financial system.

In a November 3, 2016 update, Goode relayed intelligence from his sources that the Earth Alliance, which includes operatives from the Navy and the FBI, had indeed supported the Trump campaign. Among the reasons given was that the Alliance perceived Trump as someone who would avoid embroiling the US in wars with major nations that could result in global calamity:

TRUMP IS BEING BACKED BY THE ALLIANCE

> ... Donald Trump is indeed being backed by the Alliance. Trump is certainly not entering into this battle free from his own baggage. He has said and done a variety of things that are upsetting to people. He does not, however, represent a group that actively seeks to kill billions of people by starting World War III as soon as possible.[321]

Previously, Goode said that both the Secret Space Program Alliance and the Earth Alliance would use document releases as their preferred method of ensuring policy changes, and for disclosing the truth about secret space programs. In response to a question regarding this, Goode said:

> A full disclosure event would consist of a major data dump on the Internet with many hundreds of thousands of document, audio and video files on multiple mirrored sites for everyone to have access to. There would be a collapse of the corporate media machine and a 24/7 television and radio education campaign would be initiated. Not all channels and stations would be co-opted so people are not overwhelmed and could "tune out" as they needed to due to the stress.[322]

Consequently, it is highly plausible that some of the documents that came through Wikileaks were orchestrated by the Earth Alliance, as Goode contends.

This raises the question of whether support for Trump was also due to him being identified by Navy/Solar Warden officials as someone representing a rare and valuable opportunity to them. In Trump, did they see a man who would be willing to dedicate his Presidential administration to creating another Manhattan Project, devoted to quickly building a new generation of space battle groups? Trump's real estate background in large construction ventures undeniably predisposes him toward supporting such a grandiose vision and project. This helps explain why Tompkins says the Navy is "100% behind Trump."[323]

Goode's claim that the Earth Alliance supported Trump does add weight to the possibility that he was identified as someone who would help the Navy convert its secret space program into a publicly supported program, reminiscent of *Star Trek's* fictional "Star Fleet." Since Gene Rodenberry developed *Star Trek* in the 1960's as a result of his collaboration with Leslie Stevens, who had been informed about the Navy's plans for building secret space fleets by his father, Vice Admiral Leslie Stevens, a natural progression may be taking place. If *Star Trek* was indeed planned to be a form of "soft disclosure," then it was foreseen by the Navy that one day Solar Warden would have to be revealed and converted from a highly classified program into a publicly supported space program like NASA. This would require "full disclosure" under a President who could motivate the American public into believing in such an enormously expensive "Manhattan Project," and to then build the next generation of Solar Warden space battle groups in record time.

An important clue to Trump's favorable Navy policies of the future is his appointment of Stephen Kevin Bannon as his chief strategist, equal in status to his Chief of Staff.[324] Bannon is a retired Navy officer with experience on aircraft carrier battle groups, as described in a brief biographical description on the Breitbart website:

> Before entering the business world, Mr. Bannon was in the military. For seven years, he was a surface warfare officer in the Navy. He served in battle groups that were stationed in the Arabian Sea and in the Persian Gulf. After that, he was stationed at the Pentagon, becoming special assistant to the chief of naval operations during President Ronald Reagan's first term. [325]

Mystery surrounds Bannon's work in advising the Chief of Naval Operations, since records of his military career have not been publicly released. A clue comes from an officer who served with him, retired Rear Admiral Edward "Sonny" Masso, who said:

> He's a pretty extraordinary guy ... He's very good at multi-tasking, and he can do amazing things. He was absolutely a good sailor and naval officer.[326]

It is possible that during his Naval service, Bannon learned about the Navy's secret space program and about different extraterrestrial visitors. Indeed, as a "special assistant" to the Chief of Naval Operations, he may have even participated in or known about the classified briefings Reagan received about an extraterrestrial threat during his first term.

After January 20, 2017, President Trump will be in a position to move forward with a Navy backed full disclosure initiative that will reveal advanced technologies that truly revolutionize life as we know it. Trump's emerging relationship with, and support for, the Navy is therefore a vital ingredient for full disclosure. Yet, there is another relationship that is equally important in order for full disclosure to move forward.

The FBI, Flying Saucers, Trump & Disclosure

In chapter one, it was mentioned that exactly one week after major newspapers had reported the Roswell flying saucer crash, FBI Director, J. Edgar Hoover, handwrote on a July 15, 1947 Memorandum: ""We must insist upon full access to disks recovered."[327] The newspaper articles were generated by an official US Army Air Force Press Release, made public on the morning of July 8, which was authorized by the Base Commander of Roswell Army Air Field, Colonel William Blanchard. By the afternoon, the initial Army Air Force Press Release was retracted by a more senior officer, General Roger Ramey, who said the object was a misidentified weather balloon. The two experienced Air Force Intelligence officers, who examined the Roswell debris, had simply got it wrong according to Ramey. Hoover's handwritten message clearly revealed that he did not accept Ramey's explanation, and wanted the FBI to be given access.

Hoover's handwritten message specifically referred to yet another earlier incident where a flying saucer had crashed in "the La case," and the Army denied access to the FBI.[328] Hoover was very likely referring to the Los Angeles Air Raid incident on February 24/25, 1942. A leaked "Majestic document" reveals that in July 1947, the Army Air Force had once again denied the FBI access to a crashed flying saucer stored at the time at Los Alamos National Laboratory.[329]

A number of Freedom of Information Act (FOIA) documents refer to the FBI's historic interest in the flying saucer/UFO phenomenon.[330] It is not very surprising that the FBI would have shown an interest in the flying saucer phenomenon, given widespread public sightings and involvement by citizens around the country. The most significant of the FOIA FBI documents is one confirming that Hoover wanted the FBI to launch a public investigation into the flying saucer phenomenon, provided the FBI was *given access* to crashed saucers by the [Army] Air Force.[331] The FBI was not given access, however, and Hoover subsequently ordered FBI agents not to work with the Air Force's public investigation into UFOs. In 1953, this Air Force investigation was renamed Project Blue Book, and it continued until 1969 when it was closed.[332]

While today the FBI is generally regarded as a domestic national crime fighting organization with minimal intelligence operations in foreign nations, this was not always the case. The FBI traces its origin to an investigative unit established on July 26, 1908 in the Justice Department, which a year later was renamed the "Bureau of Investigations." In 1935, Hoover became the first

Director of the newly renamed "Federal Bureau of Investigations." As World War II loomed, Roosevelt authorized the creation of a counterintelligence branch within the FBI to work with Army and Navy intelligence in combatting the Nazi menace. On June 26, 1939, Roosevelt released the following Presidential Directive:

> It is my desire that the investigation of all espionage, counter-espionage, and sabotage matters be controlled and handled by the Federal Bureau of Investigation of the Department of Justice, the Military Intelligence Division [MID] of the War Department, and the Office of Naval Intelligence [ONI] of the Navy Department. The Directors of these three agencies are to function as a committee to coordinate their activities.[333]

Roosevelt subsequently authorized the FBI to take responsibility for counterintelligence and covert operations in the Western Hemisphere. This is how a June 24, 1940 State Department Memorandum described the division of global intelligence, and counterintelligence responsibilities between the Army, Navy and FBI:

> The President said that he wished that the field should be divided. The FBI should be responsible for foreign intelligence work in the Western Hemisphere ... The existing Military Intelligence and Naval Intelligence branches should cover the rest of the world, as and when necessity arises.[334]

On July 1, 1940, Hoover subsequently launched the "Special Intelligence Service" as a branch of the FBI that would exclusively conduct US intelligence operations in the Western Hemisphere. Consequently, up until the creation of the Central Intelligence Group (forerunner to the CIA) in January 1946, it was the FBI that was responsible for intelligence operations in South America. Here is how the FBI website describes its Special Intelligence Service:

> By 1940, South America had become a hotbed of German intrigue. More than half-a-million German emigrants—many supporters of the Third Reich—had settled in Brazil and Argentina alone. In line with the Bureau's earlier intelligence work on threats posed by Germany, Roosevelt wanted to keep an eye on Nazi activities in our neighbors to the south. And when the U.S. joined the Allied cause in 1941, the President wanted to protect the nation from Hitler's spies and collect intelligence on Axis activities to help win the war. Over the next seven years, the FBI sent more than 340 agents and support professionals undercover into Central and South America as part of the Special Intelligence Service.... The service was gathering information and sending it back to FBI Headquarters in Washington, where it was crafted into useful intelligence for the military and others.[335]

While William Tompkins was involved in a covert Naval Intelligence program aimed at Nazi Germany, one of the things that the Navy spies reported was that the Nazis had poured personnel and resources into South America. The Nazis were setting up a major staging post for building their secret space program in Antarctica just prior to and during World War II. As the war came to an end, Nazi Germany increased the flow of personnel and equipment into South America and Antarctica. FOIA documents confirm that the FBI was monitoring Nazi operations in South America during this entire period and was aware that key Nazi leaders, such as Adolf Hitler and Martin Bormann, had been allowed to escape to Argentina.[336] Hoover was aware that Allen Dulles, from the Bern, Switzerland Headquarters of the Office of Strategic Services (a forerunner to the CIA), had negotiated secret deals with Nazi SS officials towards the end of the war, which had been approved by Presidents Roosevelt and Truman.

Figure 60 . "The FBI and Foreign Intelligence". Source: CIA Website

MEMORANDUM PREPARED BY ASSISTANT SECRETARY OF STATE BERLE JUNE 24, 1940, AND APPROVED BY THE PRESIDENT:

In the presence of General Sherman Miles, I telephoned the President. Referring to the conversations we have had with Mr. Welles, I said that the Inter-Departmental Committee charged with coordinating intelligence work wished his direction as to the formation of a unit for foreign intelligence work (in addition, of course, to the intelligence work now being carried on by the Army and the Navy).

The choice lay between the Federal Bureau of Investigation, the Military Intelligence Division of the Army, and the Office of Naval Intelligence.

The President said that he wished that the field should be divided. The FBI should be responsible for foreign intelligence work in the Western Hemisphere, on the request of the State Department. The existing Military Intelligence and Naval Intelligence branches should cover the rest of the world, as and when necessity arises.

It was understood that the proposed additional foreign intelligence work should not supersede any existing work now being done; and that the FBI might be called in by the State Department for special assignments outside the American Hemisphere, under special circumstances. Aside from this, intelligence outside the American Hemisphere is to be left to the officers of the Army and Navy.

Importantly, during this entire period, the FBI was primarily responsible for intelligence operations in South America, and directly investigated what the Nazi's were up to. This meant that the Navy and FBI, up to 1946, were pooling their intelligence resources to understand the extent of Nazi operations in South America/Antarctica; to determine how far Nazi flying saucer technologies had developed, and in response, develop a coordinated strategy for opposing the Nazi threat in the Western Hemisphere.

Hoover was staunchly opposed to the creation of the Central Intelligence Group, and its successor the CIA, which was established in September 1947. He refused to share documents and resources with the CIA. This is exemplified in a declassified CIA document showing that the FBI withdrew its field agents and resources from Latin America in August 1946, *before* Central Intelligence Group agents could be deployed.[337] Multiple researchers have confirmed the subsequent quarter century-long animosity Hoover had towards the CIA, which lasted up to his death in 1972. According to historian, Mark Riebling: "there have been no fewer than twelve White House initiatives to defuse the interagency conflict. All have failed ..."[338]

Many researchers assume that the FBI-CIA conflict was due to Hoover losing the bureaucratic turf war over the FBI playing a role in foreign intelligence operations, or the intrinsic challenge in harmonizing intelligence operations with law enforcement.[339] Was the real reason behind Hoover's animosity caused by his fear of the threat he knew the CIA posed, due to their efforts in establishing a close working relationship with Antarctica-based Nazis and their Reptilian allies? This became clearer as Dulles continued to conduct secret negotiations between different US administrations and Nazi leaders in South America and Antarctica, and rise through the CIA ranks to become its first civilian Director in 1953.

A close relationship was developed between the USAF, the RAND Corporation and the Nazi/Reptilian alliance after an agreement had been reached with the Eisenhower administration in early 1955. This USAF/RAND/Reptilian Alliance was opposed to the Navy, and the Nordic extraterrestrials covertly working as allies through third parties such as the Douglas Aircraft Company, which Tompkins has described at length. Hoover saw that the CIA had been infiltrated by the Nazi/Reptilian Alliance, and had become an integral part of what can be described as a USAF/CIA/Nazi/Reptilian Alliance. In contrast, Hoover's FBI would quietly assist the Navy in its covert relationship with Nordic extraterrestrials in developing future space battle groups, to prevent infiltration and sabotage by the USAF/CIA/Nazi/Reptilian Alliance.

Hoover remained FBI Director up to his death in 1972. His knowledge about flying saucers, a Nazi Space Program in Antarctica/South America, the USAF/CIA /Nazi/Reptilian alliance, and the historic cooperation between the Office of Naval

Intelligence and Nordic extraterrestrials, has become part of the institutional memory of the FBI. Once again, a television show, *The X-Files*, would become a form of "soft disclosure," opening a small window into the historical FBI association with the flying saucer phenomenon and the Nazi/extraterrestrial element.

Given the FBI's institutional memory about flying saucers, secret space programs, and history of covert cooperation with Naval Intelligence, the FBI's actions during the 2016 Presidential election raises an intriguing possibility. Did the FBI overtly assist Trump because he is seen as someone predisposed to publicly disclosing the truth about these controversial subjects, of poignant concern to both the FBI and the Navy?

Undoubtedly, the most surprising event preceding the election was an October 28 letter by FBI Director Comey stating that Clinton was once again under federal investigation.[340] The FBI had learned from a "sexting" investigation of disgraced congressman, Anthony Weiner, that one of his computers was shared with his estranged wife, Huma Abedin, deputy chair of the Clinton campaign. It contained emails between Abedin, Clinton and others sent during her tenure as Secretary of State. This was relevant to the FBI investigation of Clinton, which had been closed earlier in July 2016 when Comey recommended that no charges should be brought against Clinton.[341] Comey's letter to the heads of different Congressional Committees, informing them of the re-opening, rocked the Clinton campaign and led to 11 days of the worst press experienced by Clinton during the entire campaign. Then, on November 6, only two days before the election, Comey once again called off the investigation and lifted a cloud of uncertainty over the Clinton campaign.[342] Her supporters expressed a huge sigh of relief, and believed the road to the White House was wide open once again.

Most political pundits and media organizations were bewildered by Comey's puzzling behavior. Why did he say anything in the first place if there was nothing sufficiently incriminating in the Weiner/Abedin emails to change the earlier July decision to close the investigation? Clinton supporters were angry at Comey, and believed he had blatantly interfered in the election to help the Trump campaign.[343] Republicans, including Trump himself, criticized Comey for once again protecting Clinton prior to the election.[344]

After her election loss, Clinton told donors in a conference call that Comey's actions had indeed done her campaign irreparable harm and directly assisted Trump. She was quoted to have said that Comey's first letter on October 28 had badly affected her campaign's momentum:

> There are lots of reasons why an election like this is not successful ...our analysis is that Comey's letter raising doubts that were groundless, baseless, proven to be, stopped our momentum." [345]

According to Clinton, Comey's second letter on November 6 was even more damaging:

> Mrs. Clinton said a second letter from Mr. Comey, clearing her once again, which came two days before Election Day, had been even more damaging. In that letter, Mr. Comey said an examination of a new trove of emails, which had been found on the computer of Anthony D. Weiner ... had not caused him to change his earlier conclusion that Mrs. Clinton should face no charges over her handling of classified information. Her campaign said the seemingly positive outcome had only hurt it with voters who did not trust Mrs. Clinton and were receptive to Mr. Trump's claims of a "rigged system."[346]

Clinton's analysis correctly identifies the damage done to her campaign by Comey's 11th hour intervention. The retiring Democratic leader in the Senate, Harry Reid, went further to claim that Comey was actually a Republican Party operative: "there is no question in my mind she would have won this election without any problem if Comey had not been the Republican operative that he is." [347] Clinton and Reid's assessment raises two questions. Was Comey's intervention an intentional act designed to help Trump's campaign? Also, was Comey's intervention done in concert with other powerful institutional actors covertly helping the Trump campaign?

As mentioned earlier, Goode maintained that the "Earth Alliance" had backed Trump and was behind the Wikileaks email releases damaging to the Clinton campaign. The Earth Alliance allegedly includes senior figures in both the Navy and FBI, and also the Russian Federation. The possibility that Russia and the FBI were actively colluding in undermining the Clinton campaign was raised by Democratic critics of Comey's actions. This included the former Chair of the National Democratic Committee, Howard Dean, who tweeted: "Comey put himself on the same side as Putin." [348]

Significantly, Corey Goode said that his sources had confirmed that the FBI was indeed working with Russia, behind the scenes. Commenting on this relationship in regard to Comey's intervention that effectively sabotaged the Clinton campaign, Goode stated, "This has led to surprising public declarations of collusion between Russia and the FBI—something that we have been revealing for years now as a key aspect of the Alliance."[349] If accurate, this is an extraordinary revelation. The FBI actively colluded with both domestic and international actors to help elect Donald Trump to the US Presidency! To understand why, we need to examine the emerging relationship between Trump and President Putin, and what this potentially means for world peace, and "official disclosure" of secret space programs and extraterrestrial life.

Russia's Role in Disclosure & the Emerging Trump-Putin Relationship

In the previous chapter, the Clinton campaign was shown to be connected to a "limited disclosure" initiative. This involved framing the Russian Federation as a long term strategic enemy, when it came to a secret space program run by the Air Force. If Clinton had won the election, many political analysts expected increased tensions with Russia, which could have easily escalated into a nuclear war.[350] Provokingly, after Russia's annexation of Crimea, Clinton referred to Putin as another Hitler, thereby setting the scene for World War III.[351] In contrast, Trump promised a new approach to Russia where there would be cooperation in mutual areas of interest.

This leads to the question of whether President Trump will consider full disclosure of multiple secret space programs and extraterrestrial life a mutual area of interest with Russia? There is good reason to believe that a Trump administration will partner with Russia in an "official disclosure" initiative that will forever change life on our planet.

First, let us consider what Trump had to say about Russia both prior to and during his Presidential campaign. In a 2007 interview on *Larry King Live* on CNN, Trump expressed his admiration for Vladimir Putin's leadership in Russia:

> Look at Putin -- what he's doing with Russia -- I mean, you know, what's going on over there. I mean this guy has done -- whether you like him or don't like him -- he's doing a great job in rebuilding the image of Russia and also rebuilding Russia period.[352]

The most significant remark made by Trump during the 2016 Presidential campaign came in response to a comment by Putin on ABC News:

> [Trump is] a very colorful person. Talent, with[out] any doubt. But it's not our affair to determine his worthiness - that's up to the United States, but he is absolutely the leader in the presidential race. He wants to move to a different level of relations, to more solid, deeper relations with Russia and how can Russia not welcome that – we welcome that."[353]

Afterwards, Trump made a statement to ABC News where he again articulated his admiration of Putin, and expressed some of his future hopes concerning Russia:

> It is always a great honor to be so nicely complimented by a man so highly respected within his own country and beyond. I have always felt that Russia and the United States should be able to work well with each other towards defeating terrorism and restoring world peace, not to mention trade and all of the other benefits derived from mutual respect.[354]

Trump's reference to working with Russia in "defeating terrorism and restoring world peace" on a basis of "mutual respect" is noteworthy. It is a welcome departure from the tension and antagonism that existed at the time due to events in Ukraine, which led to the US and European Union imposing sanctions on Russia, and in turn being hit by Russian counter-sanctions.

A week after Trump's Presidential victory, Trump and Putin spoke by telephone and agreed that relations between the US and Russia were "extremely unsatisfactory," as described in a Kremlin report:

> During the call, the two leaders discussed a range of issues including the threats and challenges facing the United States and Russia, strategic economic issues and the historical U.S.-Russia relationship that dates back over 200 years," it said.

> In its readout, the Kremlin added that both Putin and Trump agreed that the U.S.-Russian ties are in "extremely unsatisfactory" condition now. "They spoke for active joint work to normalize ties and engage in constructive cooperation on a broad range of issues," it said, adding that Putin and Trump emphasized the need to develop trade and economic cooperation to give a strong basis to U.S.-Russia relations. Putin and Trump also agreed on the need to combine efforts in the fight against their No. 1 enemy – "international terrorism and extremism".[355]

A genuine relationship of mutual respect between Trump and Putin could lead to levels of cooperation between the US and Russia, not seen since the days of Ronald Reagan and Mikhail Gorbachev during the 1980's. The Reagan-Gorbachev relationship led to the end of the Cold War super power confrontation, and ushered in a period of unprecedented cooperation between the US and Soviet Union/Russia in global affairs.

A Trump-Putin relationship is likely to be just as significant, if not more so, than the Reagan-Gorbachev relationship in influencing global affairs. Reagan and Gorbachev ended the Cold War, which pitted two ideologically opposed global systems against one another—Marxism Leninism and Pluralist Democracies. Trump and Putin will soon be in a position to end another ideological divide between two global systems. This involves openly sharing the benefits of highly advanced technologies secretly developed by a transnational military-industrial complex for the exclusive benefit of global elites and national security programs.

On the one hand, there is a global system comprising highly classified "closed source" advanced technology projects, involving deep space operations using exotic propulsion systems such as antigravity, which routinely interacts with visiting extraterrestrial life. Both the Navy and Air Force have built secret space programs using these exotic technologies, and use incredibly advanced medical technologies such as age-regression, and holographic healing that can restore limbs and organs.

On the other hand, there is a global system which is "open source" and far less technologically advanced, where antigravity propulsion is considered by "open source" scientists to be a myth, extraterrestrial life is officially yet to be discovered, and age-regression technologies are deemed decades away in development to be applicable to humans.[356] This is the world known to the vast majority of our planet's population of over seven billion. The movie *Elysium* represents the "soft disclosure" of the technological gulf between two portions of humanity, and shows that this divide was allowed to grow over more than six decades because a classified research and development system was maintained. This is our own technological apartheid on a global scale.

There is good reason to believe that Putin is the most comprehensively briefed international leader when it comes to the history, development and operations of the shadow world of advanced technology programs. He has been either President or Prime Minister of Russia since August 1999. Prior to that, he briefly headed Russia's Federal Security Service, successor to the KGB, under the Boris Yeltsin Presidency from July 1998 to March 1999, and subsequently headed Yeltsin's Security Council (equivalent to the US National Security Council) before becoming Prime Minister for the first time. He replaced Yeltsin as President on May 7, 2000.

Confirmation that Putin, while President of Russia, was given briefings about extraterrestrial life comes from no less than a former Russian President, and current Prime Minister, Dmitry Medvedev. In a hot microphone incident in December 2012, Medvedev was recorded as saying in response to a question:

> Along with the briefcase with nuclear codes, the president of the country is given a special 'top secret' folder. This folder in its entirety contains information about aliens who visited our planet. Along with this, you are given a report of the absolutely secret special service that exercises control

> over aliens on the territory of our country ... More detailed information on this topic you can get from a well-known movie called 'Men In Black' ... I will not tell you how many of them are among us because it may cause panic.[357]

While many in the mass media assumed Medvedev was joking, the context of the conversation and expert analysis of his body language confirms that he was being very serious.[358] His comments help confirm that Russian Presidents receive classified briefing documents about extraterrestrials living among us, and the existence of a covert program set up to monitor aliens in Russia and elsewhere.

Confirmation that Putin is a key global leader when it comes to the operations of secret space programs comes from Goode. He says that Putin has been involved in secret meetings and negotiations, conducted by different secret space programs and extraterrestrial alliances, over the "extent and pace" in which full disclosure should occur. Goode gives the example of Putin's unexplained ten day disappearance, from March 6 to 16, 2015, as a probable case of him and/or his representatives attending a secret space program meeting held at a secret Moon base (Lunar Operations Command – LOC) to conduct disclosure negotiations:

> The timing of his 10 day disappearance occurred exactly during the time of the SSP Alliance/Sphere Alliance Conference that was taking place on the LOC along with many Earth Politicians (Non Cabal) and Regular Earth Citizens. I did not personally see him or have I been briefed that he was there. I was not involved in that portion of the conference. I do strongly suspect that he or his representatives were present.[359]

What is even more intriguing is growing evidence that Russia is actively cooperating with the *same Nordic extraterrestrials* that

Tompkins claims have been covertly assisting the US Navy since at least the 1950's!

This is a view that Dr. Preston James, a regular writer for the popular website, *Veteran's Today,* says he was also told independently by his own insider sources. James refers to a Russian agreement with a positive group of extraterrestrials (Nordics) opposed to the Draconian extraterrestrials, and what he describes as the latter's "Rothschild Khazarian Mafia" (RKM)/Cabal partners:

> Sources deep within Russia have reported that the Russian Federation has signed a treaty with a certain Alien ET group, and this particular Alien ET group has a long conflict with the a certain group that has allegedly been "advising and running" the top RKM Policy-makers who have served as their agent for taking over the whole world.[360]

James goes on to say that the Russians have been given advanced weapons technologies by the positive extraterrestrial group, which can neutralize the best technologies available to the CIA/Mossad/Saudi equipped Islamic State groups, who are backed by the Cabal/RKM:

> At this time it is unknown what their long term agenda is, but so far they have equipped the Russian Federation with some astoundingly powerful weapons, especially ultra-high tech electronics which can temporarily shut off the electrical systems of whole ships or carriers and even aircraft and all group, air and satellite radar systems. Putin has allegedly been told that he should go ahead and checkmate the RKM and its Terrorists cutouts in Syria and Iran and not worry, because this Alien ET group now has his back.[361]

James refers to an incident involving the U.S.S. Donald Cook naval vessel as an example of these advanced extraterrestrial technologies being successfully developed and deployed by the Russian military:

> Some of Putin's secret space war weapons are shocking in their actual capabilities. One was tested on an America ship, the USS Donald Cook, last April 2014. As two Russian fighters flew over, the ship's electric supply was disrupted, shutting down all radar and Aegis defense systems. It was reported that the top officers were so upset that some of them resigned their commissions and left the Navy, feeling they could not even defend their own ships and crews anymore against such ultra high-tech weapons.[362]

James information dovetails with Goode's testimony about Putin's connection to an "Earth Based Alliance" associated with the BRICS nations, which is dedicated to overturning the plans of the Cabal/RKM. Goode describes Putin's role as follows:

> Putin and other elements of the "Earth Based Alliance" that make up the BRICS Alliance among others that dove tail together are all working for the common goal of defeating the "Satanic/Luciferian Cabal" that is now in control of the majority of the world and responsible for not only amazing deceptions of his and other countries populations but also some of the most horrific Crimes Against Humanity that have ever taken place in known history. There are many of these crimes that have become more and more known through recent disclosures behind the scenes. This has only caused these groups and people to want to bring down these Secret Earth Government Syndicates now more than ever.[363]

Regarding the Donald Cook incident, Goode replied to the following email question I sent him:

> [Salla] On April 12, 2014, the USS Donald Cook equipped with an Aegis Combat System was allegedly disabled by a Russian Su-24 tactical bomber deploying a sophisticated

electronic jamming system. Was this an example of the kind of technological assistance given to Solar Warden and its earth allies to change the military/political balance of power on Earth?

[Goode] I have been told this is the case by several SSP Alliance members, but I have not seen any briefing reports personally to be able to vouch for that 100% myself. Other insiders have reported this as well so it is probably a safe speculation.[364]

Yet more evidence of an alliance between the Russian Federation and extraterrestrials emerged in an extraordinary television interview, held in May 2010, with the retiring Governor of the Russian Republic of Kalmykia, Kirsan Ilyumzhinov, who had been President of the World Chess Federation since 1995. He claimed that in 1997, he was taken from his penthouse apartment and brought on board an extraterrestrial vehicle. Ilyumzhinov stated that the extraterrestrials he met were humanoid and gave him a tour of their ship. According to Ilyumzhinov, his experience was backed by three witnesses, who had searched for him at his home after he had boarded the alien spaceship.[365]

An enormously telling fact is that Ilyumzhinov appeared on Russia's no.1 rated television station, Channel One, which is 51% controlled by the Russian government, to openly discuss the incident. The interviewer, Vladimir Pozner, began the segment with questions about Ilyumzhinov's experience. Clearly, the host and producers knew in advance of what had happened to Ilyumzhinov, and wished to discuss it on air. Also extraordinary, there was no censorship of Ilyumzhinov's experience, which was immediately made available on the Channel One website.[366] This airing of Ilyumzhinov's experience signaled a remarkable covert attempt by the Russian government to prepare its citizens for the eventual public disclosure of extraterrestrial life interacting with high level political officials.

In the interview, Ilyumzhinov described how he was awakened from his sleep, and exited his apartment through a balcony onto the waiting ship. He said:

> In the evening I read a book, watched TV, and went to rest. And then, probably, fell asleep, and felt that the balcony opened and someone called. He came up, I looked – a kind of translucent half-pipe. I went into this tube and saw people in yellow spacesuits.[367]

Ilyumzhinov went on to describe how the extraterrestrials took him on a tour of their vehicle. Then they explained that they needed samples from another planet and took him with them before returning him safely to his apartment. Before the interview was over, Ilyumzhinov shared his conclusions about the characteristics, behavior and goals of the extraterrestrials: "They are people like us. They have the same mind, the same vision. I talked with them. I understand we are not alone in this whole world. We are not unique."[368]

The way Ilyumzhinov was able to publicly share his experience on a Government controlled television station without any censorship demonstrates the tacit approval of senior Russian officials, especially Putin, who was Prime Minister at the time. Ilyumzhinov's status as both a Governor and President of the prestigious World Chess Federation ensured that his testimony would get significant public attention both in Russia and beyond.[369] At the very least, Russia was displaying what would seem to be *an extraordinary degree of public openness* on UFOs and alien life, by allowing an elected official to share his experience in this manner. Perhaps most significantly, the interview suggests that Russia was already taking serious steps with its citizenry towards official disclosure of the existence of advanced extraterrestrial life, and the high level meetings taking place between them and Russian officials.

What the above evidence suggests is that Russia's President Putin is not only aware of extraterrestrial life, but has formally entered into an agreement with the Nordics. This places Russia, the US Navy and the FBI in the same strategic alliance with Nordic extraterrestrials. In contrast, the USAF, CIA and much of the US military industrial complex is on the other side of the strategic ledger, due to secret agreements and active cooperation with the Nazi Reptilian Alliance.

Given the Navy/Solar Warden and FBI support for President Trump, it is very likely that he will eventually be informed of these different extraterrestrial groups and their multi-faceted alliances. This will subsequently lead to the development of a partnership with President Putin for an "official disclosure" that goes well beyond the "limited disclosure" initiatives, like the one from DeLonge, his advisors, and the US Air Force. Consequently, the relationships Trump develops with the Navy, FBI and Russia, can lay the foundations for "full disclosure" of Secret Space Programs and extraterrestrial life.

Endnotes

[312] Hillary Clinton says that Comey's intervention in the Presidential Election was what led to her surprise defeat, https://sputniknews.com/politics/201611131047396476-clinton-blames-comey/
[313] "Trump Outlines Plan To Build 350 Ship Navy And Revitalize America's Infrastructure," https://www.donaldjtrump.com/press-releases/donald-j.-trump-outlines-plan-to-build-the-350-ships-our-navy-needs
[314] "MEMO: Trump Announces Nationwide Ship-Building Plan To Create 350 Ship Navy," https://www.donaldjtrump.com/press-releases/trump-announces-nationwideship-building-plan-to-create-350-ship-navy
[315] "Open Letter from Military Leaders," https://assets.donaldjtrump.com/MILITARY_LEADERS_LETTER.pdf
[316] Unpublished Recorded Interview, February 25, 2016
[317] For discussion of the deep black budget, see Michael Salla, "The Black Budget Report: An Investigation into the CIA's 'Black Budget' and the Second Manhattan Project," http://exopolitics.org/Report-Black-Budget.htm
[318] "Department of Defense (DoD) Releases Fiscal Year 2017 President's Budget Proposal," http://www.defense.gov/News/News-Releases/News-Release-View/Article/652687/department-of-defense-dod-releases-fiscal-year-2017-presidents-budget-proposal
[319] See Tierney McAfee, "Hillary Clinton Says WikiLeaks Hack is Russia Trying to Mess With US Election," http://people.com/politics/hillary-clinton-wikileaks-hack-debate-russia/
[320] Corey Goode revealed that the Earth Alliance was behind the Wikileaks in a private skype communication on November 13, 2016.
[321] Corey Goode, "Are We Navigating To Our Optimal Temporal Reality?" http://spherebeingalliance.com/blog/are-we-navigating-to-our-optimal-temporal-reality.html
[322] Corey Goode, FAQ, http://spherebeingalliance.com/faqs/461-hi-corey-you-commented-on-dr-sallas-page-that-anything-less-than-full-disclosure
[323] Phone interview with Bill Tompkins, February 16, 2017.
[324] Alex Swoyer, "Trump Names Steve Bannon as White House Chief Strategist and Reince Priebus as Chief of Staff," http://www.breitbart.com/2016-presidential-race/2016/11/13/trump-names-steve-bannon-as-white-house-chief-strategist-and-reince-priebus-as-chief-of-staff/

[325] "NYT: Meet Stephen K. Bannon — U.S. Naval Officer, Harvard MBA, Investment Banker, Filmmaker, Media Mogul, Populist 'Establishment Outsider'," http://tinyurl.com/jxlyooz

[326] Mark Faram, "Trump's controversial new adviser promoted conservatism even in the Navy," http://www.militarytimes.com/articles/trumps-controversial-new-adviser-promoted-conservatism-even-in-the-navy

[327] Document available online at: http://aboutfacts.net/ufo/UFO43/Small/HooverUFO.jpg

[328] Document available online at: http://www.unacknowledged.info/j-edgar-hoover-ufo-memo/

[329] See Robert Wood and Ryan S. Wood, "Interplanetary Phenomenon Unit Summary," *The Majestic Documents* (Wood and Wood Enterprises, 1998) p. 38. Also available online at: http://www.majesticdocuments.com/pdf/ipu_report.pdf

[330] FBI, "UFOs or No? The Guy Hottel Memo," https://www.fbi.gov/news/stories/ufos-and-the-guy-hottel-memo

[331] "J. Edgar Hoover UFO Memo. The Army Retrieved Crashed Disc," http://www.unacknowledged.info/j-edgar-hoover-ufo-memo/

[332] For documents about the history of Project Blue Book, go to National Archives, "Project BLUE BOOK - Unidentified Flying Objects," https://www.archives.gov/research/military/air-force/ufos.html

[333] G. Gregg Webb, "New Insights into J. Edgar Hoover's Role: The FBI and Foreign Intelligence, https://www.cia.gov/library/center-for-the-study-of-intelligence/csi-publications/csi-studies/studies/vol48no1/article05.html#fn5

[334] G. Gregg Webb, "New Insights into J. Edgar Hoover's Role: The FBI and Foreign Intelligence, https://www.cia.gov/library/center-for-the-study-of-intelligence/csi-publications/csi-studies/studies/vol48no1/article05.html#fn5

[335] FBI, "World War, Cold War, 1939-1953," https://www.fbi.gov/history/brief-history/world-war-cold-war

[336] See Harry Cooper, *Hitler in Argentina: The Documented Truth of Hitler's Escape from Berlin, The Hitler Escape Trilogy,* (Createspace, 2014)

[337] See "Minutes of Meeting held in Room, 214, Department of State Building on Wednesday 7 August 1946, https://www.cia.gov/library/readingroom/document/cia-rdp10-01569r000100060001-4

[338] Mark Riebling, *Wedge: From Pearl Harbor to 9/11: How the Secret War Between the FBI and CIA had Endangered National Security* (Simon and Schuster, 1994), Afterword.

[339] Riebling believes this conflict is due to the basic incompatibility between intelligence operations and law enforcement, see *Wedge: From Pearl Harbor to 9/11: How the Secret War Between the FBI and CIA had Endangered National Security.*

[340] Jeff Stein, Libby Nelson, and Andrew Prokop, "New FBI letter on Hillary Clinton email investigation: What we know," http://www.vox.com/2016/10/28/13458382/fbi-hillary-clinton
[341] James Comey's July statement on the FBI Investigation into Hillary is available online at: https://www.fbi.gov/news/pressrel/press-releases/statement-by-fbi-director-james-b-comey-on-the-investigation-of-secretary-hillary-clinton2019s-use-of-a-personal-e-mail-system
[342] Julie Pace, Lisa Lerer and Jill Colvin "FBI clears Hillary Clinton in email case in last-minute reprieve," http://www.northjersey.com/news/fbi-clears-hillary-clinton-in-email-case-in-last-minute-reprieve-1.1688415
[343] Chris Strohm and Nafeesa Syeed, "FBI Shocker on Clinton Fuels Criticism of Comey's Tactics," Bloomberg, http://www.bloomberg.com/politics/articles/2016-10-29/fbi-shocker-on-clinton-probe-fuels-criticism-of-comey-s-tactics
[344] Alex Christoforou, *The Duran*, "Trump blasts James Comey's decision to clear Hillary Clinton, "it's a rigged system, and she's protected" [Video] http://theduran.com/trump-blasts-james-comeys-decision-clear-hillary-clinton-rigged-system-shes-protected-video/
[345] Amy Chozick, *New York Times*, "Hillary Clinton Blames F.B.I. Director for Election Loss," http://www.nytimes.com/2016/11/13/us/politics/hillary-clinton-james-comey.html
[346] Amy Chozick, *New York Times*, "Hillary Clinton Blames F.B.I. Director for Election Loss," http://www.nytimes.com/2016/11/13/us/politics/hillary-clinton-james-comey.html
[347] Nikita Vladimirov, NPR, "Reid: Clinton lost because of 'Republican operative' James Comey https://origin-nyi.thehill.com/homenews/news/306698-reid-clinton-lost-because-of-republican-operative-james-comey
[348] John R. Schindler, "McCarthyism 2.0 Has Infected the Democrats: Detecting nefarious Kremlin plots lurking behind every Republican bush is dangerous for democracy" http://observer.com/2016/11/mccarthyism-2-0-has-infected-the-democrats/
[349] Corey Goode, "Are We Navigating To Our Optimal Temporal Reality? Wanderers, Elections, Super Soldiers and Our Collective Consciousness," http://spherebeingalliance.com/blog/are-we-navigating-to-our-optimal-temporal-reality.html
[350] Michael Sainato, "Could a Hillary Clinton Presidency Lead to War With Russia?" http://www.truth-out.org/speakout/item/37014-could-a-hillary-clinton-presidency-lead-to-war-with-russia

---

[351] *The Guardian*, "Hillary Clinton says Vladimir Putin's Crimea occupation echoes Hitler," https://www.theguardian.com/world/2014/mar/06/hillary-clinton-says-vladimir-putins-crimea-occupation-echoes-hitler
[352] Jeremy Diamond, "Timeline: Donald Trump's praise for Vladimir Putin," CNN, http://www.cnn.com/2016/07/28/politics/donald-trump-vladimir-putin-quotes/
[353] John Santucci, "Trump Says 'Great Honor' to Get Compliments from 'Highly Respected' Putin," ABC News, http://abcnews.go.com/Politics/trump-great-honor-compliments-highly-respected-putin/story?id=35829618
[354] John Santucci, "Trump Says 'Great Honor' to Get Compliments from 'Highly Respected' Putin," ABC News, http://abcnews.go.com/Politics/trump-great-honor-compliments-highly-respected-putin/story?id=35829618
[355] Associated Press, "Putin, Trump speak by phone, agree to work to improve ties," https://www.washingtonpost.com/amphtml/world/europe/putin-trump-speak-by-phone-agree-to-work-to-improve-ties/2016/11/14/a34844c6-aaad-11e6-8f19-21a1c65d2043_story.html
[356] See Michael Salla, "Age Regression used in Secret Space Programs confirmed as Scientifically Feasible," http://exopolitics.org/age-regression-used-in-secret-space-programs-confirmed-as-scientifically-feasible/
[357] The Telegraph, "Dmitry Medvedev muses on aliens and Vladimir Putin's lateness," http://www.telegraph.co.uk/news/worldnews/vladimir-putin/9731278/Dmitry-Medvedev-muses-on-aliens-and-Vladimir-Putins-lateness.html
[358] See Michael Salla, "Russian PM not joking – extraterrestrials live among us according to MIB documentary," http://exopolitics.org/russian-pm-not-joking-extraterrestrials-live-among-us-according-to-mib-documentary/
[359] Email interview with Corey Goode, "Extraterrestrial alliance helps secret space program overcome opposition to full disclosure," http://exopolitics.org/extraterrestrial-alliance-helps-secret-space-program-overcome-opposition-to-full-disclosure/
[360] Preston James, "Putin's Wild Card in Syria," http://www.veteranstoday.com/2015/10/14/putins-wild-card-in-syria/
[361] Preston James, "Putin's Wild Card in Syria," http://www.veteranstoday.com/2015/10/14/putins-wild-card-in-syria/
[362] Preston James, "Secret Space war XIII: Alien Partners tell Putin, "Don't Worry, We've Got Your Back"," http://www.veteranstoday.com/2014/03/02/secret-space-war-xiii-alien-partners-tell-putin-dont-worry-weve-got-your-back/
[363] Email interview with Corey Goode, "Extraterrestrial alliance helps secret space program overcome opposition to full disclosure," http://exopolitics.org/extraterrestrial-alliance-helps-secret-space-program-overcome-opposition-to-full-disclosure/

---

[364] Email interview with Corey Goode, "Extraterrestrial alliance helps secret space program overcome opposition to full disclosure," http://exopolitics.org/extraterrestrial-alliance-helps-secret-space-program-overcome-opposition-to-full-disclosure/
[365] See Edward Winter, "Kirsan Ilyumzhinov and Aliens," http://www.chesshistory.com/winter/extra/ilyumzhinov.html
[366] Channel 1 website, "Guest Kirsan Ilyumzhinov. Posner. Release of 27.04.2010," http://www.1tv.ru/sprojects_edition/si=5756&fi=3800
[367] Quoted in Michael Salla, "Is Russia Preparing for Extraterrestrial Disclosure," http://www.bibliotecapleyades.net/disclosure/disclosure19.htm
[368] Edward Winter, "Kirsan Ilyumzhinov and Aliens," http://www.chesshistory.com/winter/extra/ilyumzhinov.html
[369] See, Independent, "Kirsan Ilyumzhinov: 'Chess came to Earth from outer space'," http://www.independent.co.uk/news/people/profiles/kirsan-ilyumzhinov-chess-came-to-earth-from-outer-space-2085838.html

CHAPTER 12

FULL DISCLOSURE & ANNOUNCING THE ANTARCTICA DISCOVERIES

What has been established by the evidence presented in this book so far is that the Navy began work on studying, designing and developing a secret space program as a result of an intelligence gathering operation that began in Nazi Germany during World War II. According to William Tompkins, a direct participant in this surveillance operation, the Navy was successful in ultimately building its Solar Warden program, largely due to the assistance of human-looking "Nordic" extraterrestrials. This assistance occurred covertly through both infiltration into select aerospace corporations by the Nordics, and telepathic communications with "preferred human contactees," such as Tompkins, who would receive the necessary technical information for the Navy to succeed in designing and building space battle groups capable of interstellar operations.

The Nordic extraterrestrials, Tompkins states, anticipated that the Navy would eventually become a counter weight to an earlier secret space program developed by Nazi Germany in collaboration with Reptilian extraterrestrials, which has been referred to as the "Dark Fleet." Secret agreements that began in 1955 between the Eisenhower administration and succeeding Presidential administrations with the Nazi/Reptilian alliance led to the rapid expansion of the Dark Fleet, per Goode's testimony, along with the establishment of yet another space program established by a consortium of corporations, called the Interplanetary Corporate Conglomerate.

Then other major spacefaring nations began to participate in spin-off secret space programs established under the control of the United Nations, and/or under national control, as with the Russian Federation. In the case of Russia, its program also appears to be receiving covert assistance from Nordic and perhaps other friendly extraterrestrial civilizations.

The bewildering assortment of secret space programs and the extraterrestrials alliances that are associated with them is tremendously complex, and unknown not only to the general public, but to most senior political and military officials in all nations.[370] The situation began to change significantly with the "full disclosure" provided by Tompkins and Goode, which has been sanctioned by senior officials within the US Navy and Secret Space Program Alliance. While compartmentalization and "need to know" access restrict which political leaders and military officials are briefed on these programs, shifting geo-political conditions and alliances can significantly impact this briefing process.

With the 2016 election of President Trump, a rare realignment of geopolitical forces promises to significantly impact the prospects for an officially approved process of "full disclosure" of these secret space programs, and the extraterrestrial alliances that undergird them. A key question then becomes what are the prospects that President Trump will be briefed regarding these programs, and subsequently enter into a partnership with President Putin of Russia for "full disclosure." It is this intriguing possibility that can lead us significantly beyond the "limited disclosure" that DeLonge's cooperation with the USAF is sanctioned to impart through his *Sekret Machines* initiative.

Will President Donald Trump be Briefed about Solar Warden & Extraterrestrial Life?

In contrast to Russian Presidents being briefed about, and participating in secretive negotiations over extraterrestrial life and highly classified technology programs, many US Presidents since the Eisenhower Administration are routinely kept out of the loop. For example, documents and whistleblower testimonies reveal that President John F. Kennedy was denied access to classified files dealing with advanced technology programs and extraterrestrial life by the secretive MJ-12 Group. In my 2013 book, *Kennedy's Last Stand,* I present documents that show how Kennedy's attempts to gain access and control of classified UFO files were a direct factor in his assassination.[371]

Another example of US Presidents being kept out of the loop involves President Bill Clinton, who just before beginning his first term on January 20, 1993, made the following request to close family friend and lawyer, Webster Hubbell: "If I put you over there in Justice I want you to find the answer to two questions for me: One, who killed JFK. And two, are there UFOs."[372] According to Hubbell, "Clinton was dead serious."[373] Hubbell's revelations tells us that Clinton believed his national security advisors were giving him the run around on the UFO and JFK assassination topics, and that finding answers required putting his own people into key positions.[374]

Clinton assigned Hubbell to the Department of Justice on the first day of the new Presidential administration. We know that Clinton wanted to appoint Hubble to the top job there as Attorney General, but finally had to settle for Associate Attorney General. Yet, as the third highest ranking official in the Department of Justice, Hubble enjoyed impressive executive powers and security clearances to find answers to Clinton's questions. He nevertheless failed to find the requested answers. Hubbell would eventually write about Clinton's request and efforts to find answers in his memoirs, *Friends in High Places*.[375]

The lesson here is that US Presidents, especially those from the Democratic Party, are routinely excluded from learning all there is to know about the parallel world of classified "closed source" programs.[376] There are examples, however, of US Presidents being given partial briefings, which are designed to produce a particular outcome, as exemplified in the case of President Reagan being briefed about "evil aliens."

According to an alleged 1981 briefing document, Reagan was told that there were five groups of aliens visiting Earth, one of whom were extremely hostile.[377] While there is much controversy over the legitimacy of the 1981 Reagan briefing document,[378] subsequent public comments by Reagan warning about an alien threat and the need for international cooperation suggest it is either substantially accurate, or he received a very similar national security briefing in his administration.[379]

What we do know for certain is that Reagan brought up the alien threat scenario with the leader of the Soviet Union, Mikhail Gorbachev, on several occasions, and that he requested joint US-USSR action in dealing with the problem.[380] Gorbachev later recounted the discussion from their first meeting in Geneva, Switzerland in December 1985:

> At our meeting in Geneva, the U.S. President said that if the earth faced an invasion by extraterrestrials, the United States and the Soviet Union would join forces to repel such an invasion. I shall not dispute the hypothesis, though I think it's early yet to worry about such an intrusion...[381]

At their Geneva summit, Reagan and Gorbachev began a close relationship that led to a number of stunning international developments. This included their 1987 signing of the "Intermediate-Range Nuclear Forces Treaty," resulting in the removal and destruction of short range and intermediate nuclear missiles from both sides.[382] Historians attribute the *close relationship* between Reagan and Gorbachev as the indispensable ingredient for the process that led to the ending of the Cold War.[383]

The result of Reagan and Gorbachev's private discussions about cooperating on an extraterrestrial threat was the creation of an alleged United Nations Secret Space Program. According to Goode, this program is called the "Global Galactic League of Nations" and operates in interstellar space, with at least one base in an adjoining star system.[384] The end of the Cold War can be directly attributed to the creation of this United Nations space program, which incorporated all major space faring nations in the world.[385]

It is certain that President Trump, like Reagan, will receive a classified briefing of some kind about extraterrestrial life and secret space programs. The only real question is "to what extent will Trump be briefed on such programs?" Alternatively, has he already been briefed at least informally about their existence, as hinted at in his inaugural speech?

Trump Inauguration Speech Hints at Official Disclosure

In his January 20, 2017 inauguration speech challenging the vested interests that control political life in Washington D.C., President Trump spoke of a future where humanity will have full access to the kind of advanced technologies used in the secret space programs. He appeared to be hinting at the benefits official disclosure of these programs would bring to the US and the world, and that he was going to challenge the vested interests hiding these.

Back in December 2016, Trump told presidential historian Douglas Brinkley that he planned to write a short speech himself.[386] Later, Trump tweeted a picture of himself working on his speech alone, thereby indicating that its content would be his own creation.[387] It is highly likely that Trump's speech writers would have made suggestions, polished his words, and added a few rhetorical flourishes to accentuate whatever points he wanted to make, even throughout the various drafts, leading to the final product. Nevertheless, the contents of his Inauguration Speech reveal much of what Trump really thinks about the future and hopes to achieve during his presidency.

Early in his speech, Trump was clear that Washington politics only benefited a small wealthy elite, rather than the entire population:

> For too long, a small group in our nation's capital has reaped the rewards of government while the people have borne the cost. Washington flourished, but the people did not share in its wealth. Politicians prospered, but the jobs left and the factories closed. The establishment protected itself, but not the citizens of our country. Their victories have not been your victories. Their triumphs have not been your triumphs. And while they celebrated in our nation's capital, there was little to celebrate for struggling families all across our land.[388]

Trump affirmed his campaign pledge to revitalize the American manufacturing industry, thus bringing high paying jobs back. He stated his opposition to Free Trade deals which have led to many US corporations taking their manufacturing plants out of the US, only to ship their cheaply made foreign products back at a huge profit, which goes to a small group benefited by powerful Washington DC lobbyists:

> One by one, the factories shuttered and left our shores, with not even a thought about the millions and millions of American workers that were left behind. The wealth of our middle class has been ripped from their homes and then redistributed all across the world. [389]

Towards the end of his speech, Trump articulated one sentence that contained his most developed vision of the future for America and its citizenry:

> We stand at the birth of a new millennium, ready to unlock the mysteries of space, to free the earth from the miseries of disease, and to harness the energies, industries and technologies of tomorrow.[390]

Trump's use of the phrase "unlock the mysteries of space" may be just a rhetorical flourish to show his determination to rejuvenate the NASA space program and to assist the growing commercial space industry. Alternatively, it is a hint that he is aware that there is much more happening in space than what the public has been informed about.

His use of the word "unlock" infers that these mysteries are kept hidden in highly classified programs, and he now holds the keys to them. In that sense, he is sending the message that as the President and Commander in Chief, he holds all the keys and intends to use them to further his vision of the future. Once again, he was signaling his intent of confronting the "small group" hidden in Washington that benefit from keeping secrets from the public.

The next phrase used by Trump, "to free the earth from the miseries of disease" raises the possibility that he is aware of the advanced healing technologies secretly developed in classified programs. Witnesses and whistleblowers have described advanced healing technologies used in secret space programs that can regenerate limbs and organs, and cure any disease. William Tompkins, who worked at the TRW Corporation from 1967 to 1971, says this company has developed "life extension" pharmaceutical products that could cure any disease, and physically age-regress people.[391]

Trump lastly added that he planned "to harness the energies, industries and technologies of tomorrow." Again, was this a rhetorical embellishment added by a speech writer, or was he hinting at advanced technologies used in secret space programs? In discussing the "technologies of tomorrow," Trump was clearly referring to more than just bringing back conventional manufacturing industries to generate new jobs. If some of the advanced technologies used in secret space programs, such as antigravity and free energy were released, then this would revolutionize the automobile, aviation and energy industries. Tens of millions of jobs would be created in the US alone.

Trump has promised to move quickly in achieving his campaign goals, and the vision outlined in his inauguration speech. I have learned from Corey Goode that Trump has already been informally briefed about Goode's own extensive secret space program disclosures.[392] Privately, Trump has learned about the advanced space technologies that have been kept secret from the American public, and the "small group" that benefits from this closed system. As the newly installed Commander in Chief, Trump now holds the keys to unlocking these secrets, and ushering in a "new millennium" through an official full disclosure process. Strategically, the information that launches this process, in terms of public perception, will help define it. Therefore, the Trump disclosure process is likely to begin with some startling announcements about a major archeological discovery in Antarctica.

Trump & the Impending Antarctica Discovery Announcement

In a December 11, 2016 update posted on his website, Corey Goode states that he learned of excavations taking place in Antarctica from multiple insider sources. Later on, the excavations were brought up to him by a senior officer (aka "Sigmund") within a USAF led secret space program, who has led a covert mission involving multiple abductions and debriefings of Goode.[393] During one of these abductions/interrogations, Sigmund unexpectedly shared some of his knowledge about the Antarctica excavations. This information included him describing a civilization found in the ruins led by a 12-14 foot tall "Pre-Adamite" race with an elongated skull:

> He [Sigmund] stated that an extremely ancient series of cities had been discovered flash frozen deep under the ice-shelf. He confirmed that there were also many animals and "pre-Adamites" preserved in the ice.... They were all flattened/ crushed or knocked over by the event that flash froze the area. They have tons of trees/ plants and wildlife frozen in place, like they were put on pause. He described the Pre-Adamites as beings with elongated skulls, with strangely proportioned bodies that were obviously not designed for Earth's gravity and atmospheric pressure.[394]

While the discovery of the ruins date back to the first Nazi German expedition in 1939, according to Goode's sources, it is only since 2002 that excavations by archeologists and other scientists have been allowed. Significantly, where the excavations have secretly been taking place is under two thousand feet of ice below the Ross Ice Shelf. The Antarctica ruins are well hidden from any possible aerial or satellite surveillance, thus enabling the secret to be kept this long. The archeologists have allegedly prepared documentary films and academic papers, whose release will one day soon astound the scientific community.

Goode describes that three oval shaped motherships, about 30 miles in diameter, were discovered near the site revealing that the Pre-Adamites were extraterrestrial in origin, and it was determined that they had arrived on Earth about 55,000 years ago. One of the three ships has been excavated and found to have many smaller spacecraft inside. The Pre-Adamite civilization, at least that portion of it based in Antarctica, had been flash frozen in a cataclysmic event that occurred roughly 12,000 years ago.

Goode was also told by his contacts that the most advanced technologies, and the remains of Pre-Adamites themselves, have been removed from one archeological site that will be made public. Teams of archeologists have been working with what is left, and have been told to keep all else they have seen secret.

Figure 61. Elongated Skull Discovery in Bolivia

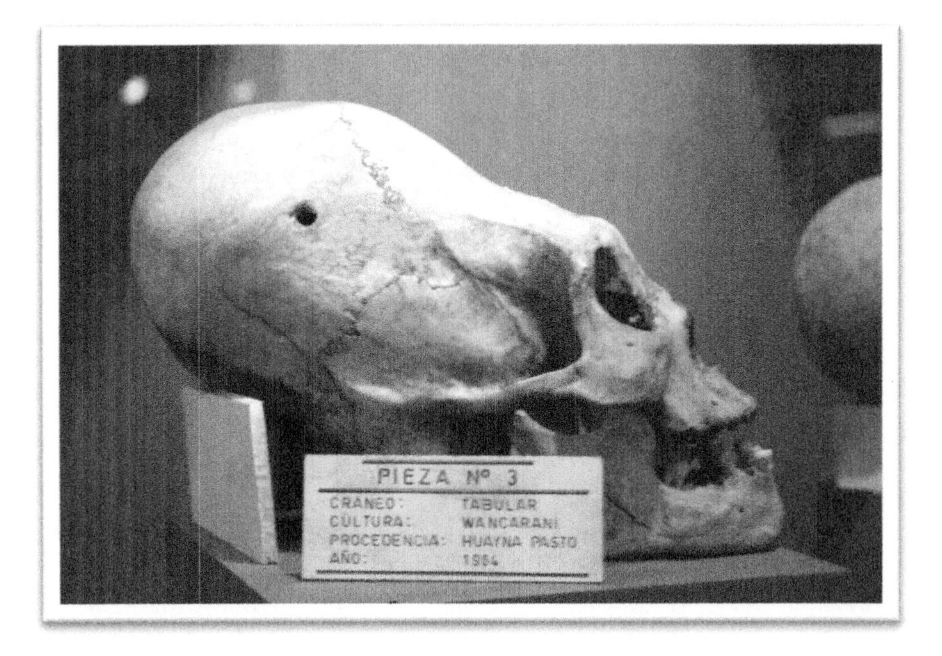

In addition, Goode's insider sources report that select ancient artifacts acquired from other locations will be brought in from vast warehouses, and seeded across the archeological site being prepared for public release. In the impending announcement about the Antarctica excavations, emphasis will be placed upon the "terrestrial elements" of the flash frozen civilization, in order not to overly shock the general population. According to Goode, the announcement is likely to be timed for distraction away from upcoming crimes against humanity trials against global elite members, because leaks and investigations will continue to emerge about international pedophile rings and child trafficking.

Up until the end of 2016, everything Goode knew about the Antarctica excavations was told to him, either by insider sources or by his very captor during an abduction. That changed in early January 2017, when Goode was taken to Antarctica to witness for himself the ruins and the excavations underway. Later that month, on January 24, I met with Goode and received an informal briefing regarding his eyewitness experience.

Goode says that shortly after New Year's Day, he was taken to Antarctica aboard an "Anshar" spacecraft. The Anshar are one of seven "Inner Earth" civilizations that Goode claims he has met with on numerous occasions. In one of his earlier reports, Goode gave the details of his experience of being taken to the main underground city belonging to the Anshar, where he witnessed their advanced technologies. He has also previously reported another experience with the Anshar, in which he was taken by them to Antarctica, where he got to see five of the working underground bases belonging to the "Interplanetary Corporate Conglomerate."[395] This corporate run secret space program is also based in this icy region.

Goode has described in other publicly shared accounts his multiple encounters with Kaaree, a High Priestess of the Anshar, who has acted as his guide and friend on many trips into the Earth's interior, Antarctica and to deep space. Another key figure in Goode's testimony is "Gonzales," who is a US Navy Lieutenant

Commander and Goode's initial contact with the Secret Space Program Alliance, which is comprised of the Navy's Solar Warden Program, along with defectors from other secret space programs. After being exposed by Goode during his involuntary abductions and interrogations by "Sigmund," Gonzales became a liaison between a Mayan Secret Space Program and the SSP Alliance, which no longer requires his presence on Earth.

In his early 2017 visit to the Antarctica ruins, Goode says he was accompanied by Kaaree, Gonzales, and two other Inner Earth Civilization representatives. One of the representatives belonged to an Asian-looking race that Goode has described in a previous account of his initial meeting with representatives from the seven Inner Earth civilizations.[396] They were taken by the Anshar spacecraft to an unexcavated portion of the ruins. This was an area that the nearby scientific teams have not yet reached, so it was still pristine and showed the full extent of a civilization that had been flash frozen.

Goode described seeing bodies twisted and contorted in various flash frozen states. The catastrophe had clearly been unanticipated. He said that the Pre-Adamites were very thin, and that it was evident from examining their bodies that they had evolved on a planet with a much lower gravitational environment. In addition to the pre-Adamites, Goode claims he also saw many different types of normal sized humans, some of whom had short tails, while others had elongated skulls similar to the Pre-Adamites. The conclusion Goode drew was that the Pre-Adamites were conducting biological experiments on the indigenous humans of the planet.

Gonzales had an instrument for taking biological samples that he plunged into the various frozen bodies. He also carried a camera and took many photos. The biological material and photos would be given to Secret Space Program Alliance scientists for study. Goode said he was unable to acquire copies of the photos for public release, at least for the time being. In addition, Goode

Figure 62. Bodies found after excavation of Ancient Pompeii

reported seeing scrolls of a metallic alloy that were rolled up with some kind of writing upon them. The Anshar and other Inner Earth representatives were collecting as many of these scrolls as possible. In earlier reports, Goode has described the Anshar Library as being quite extensive and housing many ancient artifacts from multiple civilizations.[397] Thus, the Anshar appeared to be adding the historical scroll records of this flash frozen civilization to their library.

In addition, Goode said that his party was not seen by the scientists and archeologists working on excavations in another part of the ruins. The Anshar ship had traveled through the ice to get to the ruins, and Goode related how the ship could easily move through walls using its advanced technologies.

The significance of Goode's January 2017 trip to Antarctica is that it confirms earlier briefings he had been given by various sources, including the USAF officer, Sigmund.[398] Goode's visit and confirmation of the Antarctica discovery is important in another way. It is also disturbing substantiation for the research conducted by Sir Charles Hapgood, who has studied evidence of pole shifts that have led to the Earth's axis of rotation shifting dramatically in a short period. As a result, sub-tropical areas, for example, might suddenly find themselves repositioned at the poles. His 1958 book, *Earth's Shifting Crust,* featured a foreword by Albert Einstein endorsing the rigor of Hapgood's research.[399] Hapgood summed up his theory as follows:

> Polar wandering is based on the idea that the outer shell of the earth shifts about from time to time, moving some continents toward and other continents away from the poles. Continental drift is based on the idea that the continents move individually...A few writers have suggested that perhaps continental drift causes polar wandering. This book advances the notion that polar wandering is primary and causes the displacement of continents....This book will present evidence that the last shift of the earth's crust (the lithosphere) took place in recent time, at the close of the last ice age, and that it was the cause of the improvement in climate.[400]

Hapgood's thesis that the last pole shift happened at the end of the last ice age, about 11,000 BC, would be startlingly confirmed by the discovery of a flash frozen Antarctic civilization. The flash frozen Pre-Adamite civilization is not the only case of this type of catastrophe that has impacted an ancient civilization.

The visit of many dignitaries to Antarctica in 2016, including then Secretary of State John Kerry, Apollo Astronaut Buzz Aldrin, Russian Orthodox Church Patriarch Kirill, and many other VIP's in previous years, is circumstantial evidence that a major discovery has been made in Antarctica.[401] Goode's disclosure provides testimony of the full extent of the Antarctica discovery, and the scientific excavations underway since 2002.

More support of Goode's startling claims comes from internet data mining expert, Cliff High. High describes his research method as the art of predictive linguistics, which he explains as follows:

> Predictive Linguistics is the process of using computer software to aggregate vast amounts of written text from the internet by categories delineated by emotional content of the words and using the result to make forecasts based on the emotional 'tone' changes within the larger population. A form of 'collective sub-conscious expression' is a good way to think of it. Predictive linguistics can be used to forecast trends at many different levels, from the detail of sales to individuals, all the way up to forecasts about emerging global population trends.[402]

High issues a monthly "Asymmetric Linguistic Trends Analysis Intelligence Report." In his January 2017 Report, titled "Sci-fi World," he had some very significant things to say about the global impact of a discovery in Antarctica:

> The new data sets have 'The (Antarctica) Discovery' providing more than 'new technologies' and 'economic boom period'... There is some suggestion from the data that a 'hoard' or 'trove' of 'knowledge' discovered is going to 'transform humanity' over these next '4'/four decades (and beyond)'... There are a number of sets within the Antarctica and The Discovery sets that may be describing a 'battle' that is now, and will continue to take place over 'how much', and 'if', and 'when' the 'information' about The Discovery is to be 'released' into the wilds of humanity. However, the data is quite clear that the 'arguing' and 'discord' will all be 'wasted time' as it is describing 'individuals' who just 'take it upon themselves' to 'awaken humanity' with "The Discovery."[403]

High's above analysis is very consistent with Goode's claim that the Antarctica discovery is on the verge of being publicly announced, and preparations for this have been ongoing since 2002. Indeed, Goode's account of his 2017 Antarctica visit is an example of High's reference to "individuals' who just 'take it upon themselves' to 'awaken humanity' with 'The Discovery'."

Given Trump's desire to revitalize the US manufacturing industry, it can be predicted with some confidence that he will roll out the Antarctica discovery during his administration. This will likely be a prelude to even more disclosures about advanced technologies involved in the secret space programs. The result will be, in High's words, an "economic boom period" and the release of technologies that will "transform humanity."

Peaceful social conditions are an important requirement for full disclosure to occur. Likewise, peaceful international conditions are also necessary for full disclosure, and close cooperation between Presidents Trump and Putin will go a long way in assuring this. This is where Trump's policies on issues such as Ukraine's civil war, dealing with Iran's regional power aspirations, and fairer trade relations with China will be critical. Trump's mismanagement of any of these issues can lead to violent international conflict, and set back prospects of full disclosure.

The emerging partnership between Trump, Putin, the US Navy, FBI, and "White Hats" in other institutions and nations (which Goode calls the "Earth Alliance") can create the necessary domestic and international foundations for full disclosure to occur. It is also highly significant that the full disclosure process is being covertly supported by friendly extraterrestrials, who Tompkins refers to as "Nordics" and Goode as the "Sphere Being Alliance."

Endnotes

[370] For discussion of the full range of secret space programs and their origins, see Michael Salla, *Insiders Reveal Secret Space Programs and Extraterrestrial Life* (Exopolitics Institute, 2015).
[371] See Michael Salla, *Kennedy's Last Stand: Eisenhower, UFOs, MJ-12, & JFK's Assassination* (Exopolitics Institute, 2013).
[372] Webster Hubbell, *Friends in High Places: Our Journey from Little Rock to Washington, D.C.* (William Morrow and Co., 1997).
[373] Webster Hubbell, *Friends in High Places: Our Journey from Little Rock to Washington, D.C.* (William Morrow and Co., 1997).
[374] Webster Hubbell's recollection here refutes the claim of long-time UFO researcher Grant Cameron that President Clinton was extensively briefed on the UFO/Alien issue. See Cameron's article, "UFOs – What does the President Know?" http://whitehouseufo.blogspot.com/2013/10/ufos-what-does-thepresident-know.html
[375] Webster Hubbell, *Friends in High Places: Our Journey from Little Rock to Washington, D.C.*
[376] This is contrary to the view expressed by Grant Cameron that US Presidents are given extensive briefings about the UFO/alien issue, see his article, "UFOs – What does the President Know?" http://whitehouseufo.blogspot.com/2013/10/ufos-what-does-thepresident-know.html
[377] "Transcript Of Classified Tape Recording Made At Camp David, Maryland: During A Presidential Briefing," http://www.bibliotecapleyades.net/sociopolitica/serpo/information27a.htm (accessed 8/8/15). For related article, see Steve Hammons, "Alleged Briefing to President Reagan on UFOs," http://www.bibliotecapleyades.net/exopolitica/exopolitics_reagan01.htm (accessed 8/6/15).
[378] I discuss the Reagan Briefing document and its authenticity in the in *Insiders Reveal Secret Space Programs and Extraterrestrial Alliances*, pp. 197-98.
[379] For discussion of Reagan's statements about the alien/UFO threat, see Grant Cameron, "Reagan UFO Story," http://www.presidentialufo.com/ronald-reagan/99-reagan-ufo-story
[380] For discussion of Reagan bring up the alien/UFO threat with Gorbachev, see Grant Cameron, "Reagan UFO Story," http://www.presidentialufo.com/ronald-reagan/99-reagan-ufo-story

---

[381] A. Hovni, "The Shocking Truth: Ronald Reagan's Obsession With An Alien Invasion," http://www.ufoevidence.org/documents/doc1523.htm (accessed 8/6/15).
[382] The "Intermediate-Range Nuclear Forces Treaty" was signed on December 8, 1987 and came into force on June 1, 1988.
[383] See Stanley Meisler "Reagan and Gorbachev: Warming of a Relationship," *Los Angeles Times*, http://articles.latimes.com/1988-06-02/news/mn-5847_1_president-reagan
[384] See transcript of Corey Goode interview, "Cosmic Disclosure: Global Galactic League of Nations," http://spherebeingalliance.com/blog/transcript-cosmic-disclosure-global-galactic-league-of-nations.html
[385] See Michael Salla, *Insiders Reveal Secret Space Programs and Extraterrestrial Alliances*, pp. 197-210.
[386] See Kevin Liptak, "Trump writing own 'short' inaugural speech," http://www.cnn.com/2016/12/29/politics/trump-writing-short-inauguration-speech/index.html
[387] See: https://twitter.com/realdonaldtrump/status/821772494864580614
[388] "Read Donald Trump's Full Inauguration Speech," https://www.yahoo.com/news/read-donald-trump-full-inaugural-172850356.html
[389] "Read Donald Trump's Full Inauguration Speech," https://www.yahoo.com/news/read-donald-trump-full-inaugural-172850356.html
[390] "Read Donald Trump's Full Inauguration Speech," https://www.yahoo.com/news/read-donald-trump-full-inaugural-172850356.html
[391] "Cosmic Disclosure: Validating the 20 & Back Program with William Tompkins," https://spherebeingalliance.com/blog/transcript-cosmic-disclosure-validating-the-20-and-back-program-with-william-tompkins.html
[392] Private discussion with Corey Goode on January 23, 2017.
[393] Corey Goode and David Wilcock, "Endgame Part II: The Antarctic Atlantis & Ancient Alien Ruins," https://spherebeingalliance.com/blog/endgame-part-ii-the-antarctic-atlantis-and-ancient-alien-ruins.html
[394] Corey Goode and David Wilcock, "Endgame Part II: The Antarctic Atlantis & Ancient Alien Ruins," https://spherebeingalliance.com/blog/endgame-part-ii-the-antarctic-atlantis-and-ancient-alien-ruins.html
[395] See Michael Salla, "Secret Space Programs Battle over Antarctic Skies During Global Elite Exodus," http://exopolitics.org/secret-space-programs-battle-over-antarctic-skies-during-global-elite-exodus/?forwardie=1

---

[396] See Michael Salla, "Secret Space Program Alliance Negotiates with Council of Ancient Earth Civilizations," http://exopolitics.org/secret-space-program-alliance-negotiates-with-council-of-ancient-earth-civilizations/
[397] See Michael Salla, "Sitchin's Sumerian Text Translations Contrived by Illuminati to Promote False Alien Religion," http://exopolitics.org/sitchins-sumerian-text-translations-contrived-by-illuminati-to-promote-false-alien-religion/
[398] Corey Goode and David Wilcock, "Endgame Part II: The Antarctic Atlantis & Ancient Alien Ruins," https://spherebeingalliance.com/blog/endgame-part-ii-the-antarctic-atlantis-and-ancient-alien-ruins.html
[399] Charles H. Campbell, Earth's Shifting Crust: A Key to Some Basic Problems of Earth Science (Pantheon Books, 1958). Available online at: https://archive.org/stream/eathsshiftingcru033562mbp/eathsshiftingcru033562mbp_djvu.txt
[400] Cited online at: http://www.poleshift.org/Charles_Hapgood.html
[401] For discussion of other VIP visitors to Antarctica, see Joseph Farrell, "Antarctica Update: More Strange Visitors..." https://gizadeathstar.com/2016/12/antarctica-update-strange-visitors/
[402] "About Predictive Linguistics and our Methods," https://www.halfpasthuman.com/ALTA)_how.html
[403] Cliff High's "January 2017 ALTA: Sci-fi World" report is available for purchase online at: https://www.halfpasthuman.com/Hph_reports.html

CHAPTER 13

THE ROLE OF NORDIC EXTRATERRESTRIALS IN FULL DISCLOSURE

While considering all the information presented thus far, it is important to bear in mind that both the documents provided directly by William Tompkins, and those independently released through the Freedom of Information Act, add significant credibility to his remarkable testimony. When combined with the independent witness reports by retired Navy officers, it can be concluded that Tompkins' testimony ranks very high in terms of reliability, and may be one of the most groundbreaking and informative bodies of disclosure material ever to publicly emerge from one source. What adds another layer of significance to his testimony is the considerable corroboration it lends to the earlier information revealed by Corey Goode.

It dovetails greatly with Goode's claim that humanity is being assisted by positive extraterrestrial groups that want to assist the full disclosure process. Yet in reviewing Tompkins' testimony, it is clear that the assistance he witnessed being provided by Nordic extraterrestrials was largely technical and scientific. This runs contrary to what was previously known about this specific alien group.

Beginning from the early 1950's, there has been a succession of "contactees" coming forward with incredible stories of Nordic extraterrestrials wanting to disclose their presence to the world. George Adamski, Howard Menger, George Van Tassel, Orfeo Angelucci, and Alex Collier are but a few of the many whose testimonies have inspired millions with their accounts of friendly space brothers, here to help humanity spiritually evolve to properly handle the challenges posed by the acquisition of advanced technologies now used by our modern civilization.[404] The Nordics appeared most intent about warning humanity of the dangers posed by thermonuclear weapons, which history shows were widely ignored by previous political and military leaders.

Yet during this same period, as Tompkins tells it, the Nordics were quietly and secretly helping the US Navy in the design and construction of its secret space battle groups. "Contactees" such as Tompkins were telepathically assisted in developing designs and prototypes for advanced aerospace technologies. This telepathic communication process was augmented by the Nordics infiltrating companies such as Douglas Aircraft, where they could be in close proximity to their "contactees" as these special individuals worked through the various problems that would arise in developing accurate designs and prototypes.

Why did the Nordics, if we are indeed talking about the same group of extraterrestrials, behave in such a contradictory manner? What is the reason for them warning one group of contactees about the danger of advanced technologies and promoting the disclosure of their presence through these select individuals, while helping another group develop advanced technologies in a covert manner without any public disclosure being advocated at the time?

An answer appears to arise from a broader dilemma confronting humanity, and the strategic problems this created for the Nordics. They were well aware of the aggressive Reptilian extraterrestrial group, which had imperial designs over the galaxy, and secretly reached agreements with the Nazis. With vast material help from the Reptilians, the Nazis were able to establish a breakaway civilization in Antarctica with an advanced space program, and then proceeded to infiltrate the military industrial complex of the US and other industrialized nations—with the notable exception of Russia.

In response to this complex situation, the Nordics appeared to undertake a dual track approach. On the one hand, they contacted private citizens and encouraged them to publicly reveal the existence of extraterrestrial life. On the other hand, simultaneously, the Nordics strategically identified which elements within the US military industrial complex would most likely resist the Nazi/Reptilian infiltration efforts, and eventually help the US break free of this nefarious influence.

The Nordics chose the US Navy and individuals like Tompkins, Rear Admiral Rico Botta, and Navy Secretary James Forrestal, who were all "preferred human contactees" as Tompkins put it. Forrestal was guided by the Nordics to initiate the Naval intelligence gathering program to discover what the Nazis were doing, and to put Botta in charge. Tompkins asserts that Botta reported directly to Forrestal, and thereby bypassed more senior Navy officials in the Bureau of Aeronautics, Office of Naval Intelligence and the Chief of Naval Operations.[405] This was because Botta, due to his unconventional engineering background, was more capable of appreciating the significance of the incredible information he was receiving, and able to adequately act on the dangers presented by the Reptilian alliance with the Nazis.

Especially important, according to Tompkins, was that the Nordics were very aware of Reptilian efforts to sabotage the Navy's nascent space program. The Nordics were very effective in countering this sabotage with individuals such as Tompkins, through whom they provided the necessary technical advice for the success of the Navy's massive engineering and construction efforts.

Tompkins insists that Navy officials were well aware of the covert assistance they were receiving from the Nordics, and even facilitated it by assigning prominent roles to individuals recognized as "preferred human contactees," such as Tompkins. Through Navy League programs, the Nordics continued to work with Tompkins to help prepare the next generation of Navy officers, through the Sea Cadets, about the reality of extraterrestrial life and the Navy's secret space program. Highly noteworthy, Tompkins' relationship with both the Navy and the Nordics continue to the present day.

Tompkins describes the Nordics as especially helpful in developing life extension technologies for use in the space programs. The Nordics allegedly live up to two thousand years and first helped scientists in Maria Orsic's Vril Society prior to and during the Nazi era, and later the US, in developing life extension technologies. Tompkins says that he was involved with classified life extension projects during their initial phases, while employed with TRW from 1967 to 1971, and this has resulted in significant breakthroughs. This goes well beyond their use in the "20 and back" programs described in chapter nine.

Tompkins says that pharmaceutical products have recently been developed that can restore a person, no matter their current physical age, to their late 20's for males, and early 20's for females. In addition, brain capacity can be enhanced by a factor of 400%, as he described in this interview:

> There is a study that we did later on at TRW on advanced life systems – extended life. And that program is down now to, within less than two years, it's going to be available to some people on this planet. The way it works – I'm very involved with it – essentially you take four aspirin [pills] over six months, pop them. Or you get four shots. You immediately change... What you do is you revert back to – the girl is 21 and the guy is 29. Now, it takes a while for you to do this. You then stay at that time [age] for essentially a couple of thousand years.
>
> Your brain then . . . which collectively we're only using 2.2% of our brain. I don't care what they're telling us. We're only using 2.2%. You get a minimum of 400% capability over

> what you normally had. Now, what this does is this allows you to contribute.[406]

Human genetics, Tompkins attests, contain the potential for extended life spans and higher brain capacities similar to what the Nordics normally experience, because our genetics are essentially the same. He says humanity currently does not experience such long life spans due to a combination of factors, which include harmful gasses being pumped into the atmosphere by Reptilians and their allies in order to limit human capacities and capabilities. In the following interview, he explained:

> So this then touches with the numbers of times per month that the [Reptilian] extraterrestrials drop the five gases on the industrial areas of the planet. These tankers – there's several sizes of them, some of them are only about 150 feet long. They are a very strange shape for a supersonic vehicle. Of course, they come from a mothership.... So they operate at a low altitude, not because they are afraid they're going to be seen, but because they want the gases to be distributed close to the people.[407]

Tompkins maintains that the Nordics have been covertly assisting companies such as TRW (now part of Northrup Grumman) in developing a range of pharmaceutical products, such as the life extension pills to counteract the biological effects of the toxic gasses pumped into the atmosphere by Reptilians and their allies. If Tompkins is correct, this provides an explanation for the chemtrails phenomenon, which many researchers have concluded is responsible for spreading airborne diseases such as Morgellons.[408]

According to Goode, the Nordics belong to a "Super Federation" of between 40-60 human-looking extraterrestrial races that have been intervening on Earth for hundreds of thousands of years in 22 long-term genetic experiments. It's feasible to assume that these genetic experiments are designed to test how humans respond to a range of environmental conditions, where alien interventions either inhibit or assist human evolution. Therefore, it is quite plausible that the Nordics and other human-looking aliens have been countering the efforts by Reptilians and their allies to pollute the atmosphere, water and food supply, and alter human genetics. The life extension and brain enhancement pills developed by TRW, with Nordic assistance, in this light appear to be yet another attempt to alter human DNA in order to help individuals overcome toxic pollutants and attain their fullest potential.

It would be incorrect to infer that the Nordics themselves depend on pharmaceutical products to achieve the remarkable thousand year life spans they are credited with by Tompkins, Goode and others. In his 1955 book, *Inside the Flying Saucers,* Adamski provides special insight into exactly how the Nordics are able to achieve their extended life spans. He describes being inside one of their motherships when he saw a remarkable picture whose significance was explained by his Nordic host, "Kalna":

> On the wall exactly opposite the door through which we had entered hung a portrait which I was certain must represent Deity. The emotion which the beauty of the two young [Nordic] women had aroused in me was momentarily forgotten as the wonderful radiance emanating from the portrait enveloped me. It showed the head and shoulder of a Being who could have been eighteen to twenty-five years of age, in whose face was embodied the perfect blended balance of male and female, and whose eyes held a wisdom and compassion beyond description. I do not know how long I was enrapt by this beauty. There was no interruption, until I myself returned to an awareness of my surroundings. I did not need to ask who this Being was. Kalna broke the silence by saying: "That is the symbol of Ageless Life. You will find it in every one of our ships as well as our homes. It is because we keep this symbol always before us that you will find no age amongst our people.[409]

What Adamski was told suggest that the Nordics, with their unrestricted brain capacity and unpolluted bodies, are able to hold the idea of youthful vitality in their consciousness in a way that directly influences their genetics. Put simply, aging is not experienced by them because their consciousness does not allow the idea of being old to enter. What is remarkable here is that the science of epigenetics, which is based on a similar theory proposed by biologists, such as Dr. Bruce Lipton, was developed more than seven decades after Adamski's book![410]

According to the Kardashev scale for categorizing advanced civilizations, the Nordics (as well as Reptilian and Gray extraterrestrials) fall into the categories of either Type I civilizations (able to work with planetary energies); or Type II civilizations (able to work with energies at a stellar level).[411] They need to be distinguished from a more highly evolved group of extraterrestrials Goode describes as the "Sphere Being Alliance," which are Type III extraterrestrials that work with galactic scale energies.

Goode says that the Sphere Being Alliance has created a basic technological parity, between different space programs and extraterrestrial alliances, in terms of offensive and defensive weapon technologies. This is consistent with what was discussed in chapter 11 regarding the claims by Dr. Preston James that Russia was being secretly assisted by "Nordic extraterrestrials" in developing defensive weapons. This technological parity is essential to the different factions of humanity negotiating the peaceful transition of power, enabling them to move it from a small group of global elites (the Cabal), into the hands of genuinely representative organizations that will promote the interests of all people on the planet. This peaceful transition is critical for humanity so it may adequately prepare for upcoming stellar events such as a possible "solar sneeze" involving immense coronal mass ejections that Tompkins and Goode's insider sources predict is highly probable.

Lack of preparation for such an event can be catastrophic, exemplified by what occurred to the flash-frozen civilization discovered in Antarctica. If Presidents Trump and Putin, or other world leaders do move forward, the predicted announcement of the Antarctica discovery will carry with it the implicit warning that such events have not only been regular occurrences in human history, as Charles Hapgood claimed, but may occur again in the very near future.

Goode has pointed out that the Sphere Being Alliance has strongly emphasized spiritual development and consciousness-raising (the "hippy love and peace message") as indispensable for humanity dealing with the challenges it faces.[412] At an individual level, Goode has advised people to increase their "service-to-other" activities in order to help create the optimal time-line of circumstances, and to also be prepared for impending stellar events that he has been told are part of an upcoming "Ascension Event."[413] While the concept of a "service-to-other" comes from the "Law of One" or "Ra" material, this was not the only literature recommended to Goode by officials running the secret space programs for understanding current cosmological events and the nature of consciousness. In August 2015, in his response to a question, Goode said:

> I knew that some groups were ordered to read the "RA" and "SETH" Data while in the programs at the same time I knew many groups were working to discredit both works.[414]

The *Seth* books (1963-1984) preceded the *Law of One* books (1981-1984) by almost two decades, and are widely regarded as among the finest and most influential channeled material ever to be released. While the Law of One focused on presenting a comprehensive conceptual framework for understanding how consciousness evolves, the Seth books in contrast focused on concrete steps an individual could take to change their personal and collective reality by altering their conscious belief systems.

Goode has been told a service-to-other polarity, recommended in the Law of One material, is essential for humanity at this point in our personal and collective evolution. Complimentary, the approach taken by the Seth material is to explain the nature of consciousness and how it relates to the body, allowing movement in this evolutionary direction. According to Seth:

> The body's main purpose is not only to survive but to maintain a quality of existence at certain levels, and that quality itself promotes health and fulfillment.[415]

In this regard, the ability to survive, let alone live a thousand years, becomes meaningless if the "quality of life and experience" does not meet the minimum requirements deemed by the species itself as acceptable.

One of the vast problems, in terms of individual and mass perceptions, is a sense of hopelessness or lack of personal power to change things for the better. In the Seth book, *The Individual and the Nature of Mass Events*, many illustrations are given of those who saw themselves as "effective rather than ineffective," thus rising up to do heroic deeds within otherwise mundane lives because they did not succumb to despair. Seth explains how a negative belief can be undermining:

> Despair or apathy is a biological "enemy." Social conditions, political states, economic policies, and even religious or philosophical frameworks that foster such mental states, bring about a biological retaliation. They act like fire applied to a plant.[416]

If, as Seth states, "the quality of life is important above all," then how can individuals fuel the belief in being "effective" to maintain and enhance the very quality of life, which by nature will be interconnected with others.[417] Here is Seth's answer that may surprise some:

> When you are having fun, you are helping others. When you are not having fun, and telling yourself that you are helping others, you are not helping them or yourself. So when you think in terms of responsibility, and when you make a division in your mind between responsibility and joyful fulfillment, then you are denying yourself and the world much pleasure, and hiding ... from yourself and the world the great joyful symphony that is yourself.
>
> When you are fulfilling the joyful nature of your being, you are helping yourself and you are helping others. When you help others because you think you must, but it goes against the grain, then they know it, and you inflict upon them the obligation that you have no right to inflict.[418]

Seth provides clarity on how one can achieve this positive orientation, and feel connected to the needs of others. It does not come from sacrificing one's own joy or needs to make others happy, as a service-to-other polarity might suggest to some, but by following one's own life path:

> When you follow your own nature, you automatically and naturally feel for the needs of others ... When you are joyful and free, and when you are having fun, you automatically feel ... your oneness with all other creatures of the universe, and you know your place in All That Is.[419]

Incredibly, this "joyful and free" attitude is what Tompkins says distinguished the Nordic extraterrestrials he worked with at various corporations during his aerospace career. His book, *Selected by Extraterrestrials,* highlights how fun-loving the Nordics were, despite the many dangers they faced and the complexity of situations they often were dealing with.

This meant that the Nordics, by example, were naturally showing how they had mastered a sense of "oneness" that was evident in their fun-loving attitude, reflecting the pursuit of their "own nature," inclusive of advanced technological development. Therefore, the Nordic extraterrestrials Tompkins describes meeting had achieved a rare balance of spiritual understanding along with technological development. People do not have to spend months in meditation or follow rigorous spiritual exercises to raise their consciousness, but Seth states that "consciousness requires new experience, challenge and accomplishment."[420] When people strive to connect instead of disengage, believing that their innate inner wisdom is guiding them, fun and joy can freely enhance life's quality, serving not only oneself, but others too. The advice of the famed mythologist Joseph Campbell to "follow your bliss," therefore, becomes a critical step in raising one's consciousness.[421]

Consequently, creating the optimal timeline for full disclosure and an "Ascension" event means having the courage to freely follow a true calling in life coming from within, finding the fun and joy even in challenges, and then realizing satisfying accomplishments by overcoming them. This spiritual advice is a key part of the assistance provided by the Nordics and other extraterrestrials through their "contactees" such as Tompkins. In turn, authorities in the Secret Space Program Alliance emphasize the Seth material as Goode states, and recognize its value to influence personal reality.

Conclusion

Documentary evidence provided by Tompkins in this book clearly reveal that the US Navy has been involved since the 1950's in designing and developing kilometers-long space battle groups, which it views as critical for planetary defense. The Navy's interest in building space battle groups is an outgrowth of its earlier experiments with flying aircraft carriers that date back to the 1920's, when the Navy commissioned the Goodyear and Zeppelin companies to build the USS Akron and USS Macon.

Tompkins' testimony, and the degree to which covert Navy officials have encouraged him to come forward with his information, reveal that he is part of a Navy sanctioned "full disclosure" process. Consequently, there is good reason to be optimistic that the foundations for full disclosure are currently being laid as the Trump administration appears to be coordinating with the Navy, FBI, and Russia in developing its policies. Some of these policies are focused on revealing the existence of highly classified technology programs, extraterrestrial life and even archeological discoveries, such as the site in Antarctica.

Among the many benefits of full disclosure is that it will bring an end to a decades-long separation between a technologically advanced "breakaway human society" and the rest of humanity. This technological form of apartheid is global in scope, devastating in the impact it has had on the quality of life of citizens around the world, and has spawned many criminal abuses. Furthermore, full disclosure would put a stop to any harmful practices involving toxic chemicals deliberately inserted into the atmosphere to severely limit our innate human potential. It is very doubtful that the "limited disclosure" initiative supported by the USAF would bring an end to this technological division. The same applies to those doctoring the content of artifacts discovered in Antarctica to create another limited disclosure scenario, brought on by impending announcements featuring leading scientists and archeologists to reveal only what is authorized.

Limited disclosure would most likely usher in a more subtle form of "technology apartheid," which for decades has continued maintaining the global system by which our innate capacities continue to be inhibited through toxins covertly pumped into the air, water and food supply, keeping us as an easily exploited species. Ending the artificially created global system of technology apartheid between different portions of humanity is essential for developing a global leadership that genuinely represents the aspirations of all humanity. Most importantly, full disclosure will facilitate our evolution as a species, where human life spans and mental capacities are restored. This is the innate potential we possess through our genetic connection to the Nordics and other human-looking extraterrestrial groups, thereby allowing us to join the rest of the Galactic community as full partners.

Endnotes

[404] For discussion of the early contactees, see Michael Salla, *Galactic Diplomacy: Getting to Yes with ET* (Exopolitics Institute, 2013).
[405] Phone Interview, February 16, 2017.
[406] Tompkins interviewed in "Cosmic Disclosure: SSP Testimonials with William Tompkins," http://spherebeingalliance.com/blog/transcript-cosmic-disclosure-ssp-testimonials-with-william-tompkins.html
[407] Bill Tompkins interviewed by Jeff Rense, "Our Technology Decades Ahead of What's Known," http://spherebeingalliance.com/blog/our-technology-decades-ahead-of-whats-known-part-1.html
[408] See Elena Freeland, Chemtrails, HAARP, and the Full Spectrum Dominance of Planet Earth (Feral House, 2014).
[409] George Adamski, *Inside the Flying Saucers: The Strangest Journey Ever Made!* (Warner Books, 1955) p. 49
[410] Bruce Lipton, *The Biology of Belief: unleashing the power of consciousness, matter and miracles* (Mountain of Love, 2005)
[411] For discussion of Kardashev scale see, *Futurism*, "The Kardashev Scale – Type I, II, III, IV & V Civilization," http://futurism.com/the-kardashev-scale-type-i-ii-iii-iv-v-civilization/
[412] Corey Goode, "A Good Time To Be a Cabal Defector or Whistleblower," http://spherebeingalliance.com/blog/a-good-time-to-be-a-cabal-defector-or-whistleblower.html
[413] See Corey Goode, "Are We Navigating To Our Optimal Temporal Reality?" http://spherebeingalliance.com/blog/are-we-navigating-to-our-optimal-temporal-reality.html
[414] See Corey Goode interview, "Reagan Speech about Alien Threat linked to Secret UN Interstellar Space Fleet," http://exopolitics.org/reagan-speech-about-alien-threat-linked-to-secret-un-interstellar-space-fleet/
[415] Jane Roberts, *The Individual and the Nature of Mass Events* (Amber Allen Publishing, 1995 [1981]) p. 40.
[416] Jane Roberts, *The Individual and the Nature of Mass Events*, p. 21.
[417] Jane Roberts, *The Individual and the Nature of Mass Events*, p. 24.
[418] Susan M. Watkins, *Conversations With Seth: The Story of Jane Roberts ESP Class, Combined Volumes 1 & 2* (Moment Point Press, p. 179)
[419] Susan M. Watkins, *Conversations With Seth: The Story of Jane Roberts ESP Class, Combined Volumes 1 & 2* (Moment Point Press, p. 179)
[420] Jane Roberts, *The Individual and the Nature of Mass Events*, p. 24.

---

[421] Joseph Campbell, "Quotes," http://www.goodreads.com/quotes/143093-follow-your-bliss-if-you-do-follow-your-bliss-you

INDEX

ABOUT THE AUTHOR

Dr. Michael Salla is an internationally recognized scholar in international politics, conflict resolution and U.S. foreign policy. He has held academic appointments in the School of International Service & the Center for Global Peace, American University, Washington D.C. (1996-2004); the Department of Political Science, Australian National University, Canberra, Australia (1994-96); and the Elliott School of International Affairs, George Washington University, Washington D.C. (2002). He has a Ph.D in Government from the University of Queensland, Australia. During his academic career he was the author/editor of four books focusing on international politics. Dr. Salla has conducted research and fieldwork in ethnic conflicts involving East Timor, Kosovo, Macedonia, and Sri Lanka. He has been awarded significant financial grants from the United States Institute of Peace and the Ford Foundation for peacemaking initiatives involving mid-to-high level participants from the East Timor conflict.

Dr. Salla is more popularly known as a pioneer in the development of 'exopolitics', the study of the main actors, institutions and political processes associated with extraterrestrial life. He wrote the first published book on 'exopolitics' in 2004, titled *Exopolitics: Political Implications of the Extraterrestrial Presence*, and followed this with another, *Exposing U.S. Government Policies on Extraterrestrial Life* in 2009, examining exopolitics and U.S. Foreign Policy. He authored *Galactic Diplomacy* in 2013; and this was followed by *Kennedy's Last Stand* (2013), which investigated the relationship between classified UFO's and the Kennedy Assassination. His book *Insiders Reveal Secret Space Programs and Extraterrestrial Life* (2015) investigated whistleblower testimonies on multiple classified space programs, and became an Amazon.com bestseller. He is Founder of the *Exopolitics Institute* and the *Exopolitics Journal*. Dr. Michael Salla's website is: www.exopolitics.org

Printed in Great Britain
by Amazon

41615261R00193